Theory of Linear Operators in Hilbert Space

Volume II

Theory of Linear Operators in Hilbert Space
Volume II

N. I. Akhiezer
University of Kharkov

The late
I. M. Glazman
Polytechnic Institute, Kharkov

Translated by
E. R. Dawson
University of Dundee

English translation edited by
W. N. Everitt
University of Dundee

Published in association with
Scottish Academic Press, Edinburgh by
Pitman Advanced Publishing Program
Boston · London · Melbourne

PITMAN PUBLISHING LIMITED
39 Parker Street, London WC2B 5PB

PITMAN PUBLISHING INC
1020 Plain Street, Marshfield, Massachusetts

Associated Companies
Pitman Publishing Pty Ltd, Melbourne
Pitman Publishing New Zealand Ltd, Wellington
Copp Clark Pitman, Toronto

AMS Subject Classifications: (main) 46C, 46E, 47A, 47B, 47E05, 47G05
(subsidiary) 34A30, 34B20, 34B25

Library of Congress Cataloging in Publication Data

Akhiezer, Naum Il'ich, 1901–1980
Theory of linear operators in Hilbert space.

(Monographs and Studies in Mathematics)
Revised translation of Teoriya lineĭnykh operatorov.
Bibliography: p.
Includes index.
1. Hilbert space. 2. Linear operators.
QA322.4.A3813 1980 515.7'33 80–18258
ISBN 0-273-08495-8-(v.2)

Translated from TEORIYA LINEĬNYKH OPERATOROV V GILBERTOVOM PROSTRANSTVE (TOM II) by E. R. Dawson

Third Russian edition published by Vishcha Shkola, Kharkov
English edition published by Pitman in association with Scottish Academic Press

ISBN 0 273 08496 8

Printed and bound in Great Britain
at The Pitman Press, Bath

Contents

Preface to the English Edition

The second edition of this book was published in Moscow in the summer of 1966; in May 1968 the life of my co-author and great friend, Professor I. M. Glazman, was suddenly cut short. When the question arose of translating this edition into English, I was faced with the laborious task of undertaking by myself all the work of preparing the manuscript for a new edition.

In the earliest editions of this book we tried to set out in a systematic and accessible way the fundamental concepts and facts of the theory of linear operators in Hilbert space. This is still the object of the present edition. My revision of the book has reduced itself essentially to local, albeit numerous,† amendments at various points. I have paid particular attention to the correction of inadequacies and some mistakes, and I should like to believe that there remain few essential errors in the new edition.

I take the opportunity to express my gratitude to all my colleagues and pupils who have made helpful criticisms since the publication of the second edition of this book. Invaluable assistance in the work of preparation has been afforded me by F. S. Rofe-Beketov and L. L. Vaksman: I offer them my deep gratitude.

I regard the publication of this book in Great Britain as a high honour, and I sincerely thank Professor Everitt for extending the invitation; I also thank E. R. Dawson of the Department of Mathematics in the University of Dundee for carrying out the translation, and finally I thank the publishers for their decision to publish the English translation.

Kharkov, September 1975 N. I. Akhiezer

† *Translator's note.* 170 manuscript pages of amendments and new material. My original translation was from the second Russian edition and this additional material. The third Russian edition was published in two parts in Kharkov, the first part in 1977, and the second part in 1978; this third Russian edition has been used to revise the English translation.

Preface to the Second Edition

In this book, as in its first edition which appeared fifteen years ago, the authors have set themselves the task of giving a systematic and intelligible account of the basic concepts and results in the theory of linear operators in Hilbert space. In preparing this new edition, the original text has been improved and several new sections have been added dealing with both new and classical theory. The most substantial extensions are in the exposition of the theory of completely continuous operators, of spectral analysis and the theory of rings of self-adjoint operators, and also of the theory of semi-bounded operators. A chapter on perturbations and wave operators, and an appendix on integral operators, have been added.

F. S. Rofe-Beketov has greatly assisted the authors in preparing the manuscript, has made considerable improvements in some of the proofs, and, thanks to him, a number of errors have been removed. For all this and his careful editing of the book the authors express their deep gratitude.

The authors are also very grateful to all their colleagues and students who pointed out various errors in the first edition, and to A. Z. Ryvkin for his careful attention to the manuscript during the editorial work.

Kharkov,
December 1965

N. I. Akhiezer
I. M. Glazman

While this edition was being printed, we learnt with deep regret that Professor Akhiezer died on the 3rd of June 1980.

We now dedicate this English edition to the memory of Naum Il'ich Akhiezer (6.3.1901–3.6.1980) and Izrail Markovich Glazman (21.12.1916–30.5.1968).

E. R. Dawson
W. N. Everitt

Notation

This book consists of 9 chapters and 2 appendices, but it is also subdivided into sections, marked §1 to §132. Numbering of definitions, theorems, lemmas and equations starts afresh in each section. This facilitates reference back within each section; when reference is made to a result in a previous section, the relevant section number is given as well as the number of the theorem or equation referred to in that section.

Throughout the book roman (capital) letters are used to denote spaces and point-sets; italic capital letters are generally used for functionals and operators; exceptionally gothic $\mathfrak{F}$ is used for the Fourier-Plancherel operator (§42). Bold-face **N**, **Z**, **Q**, **R**, **C** have their usual meanings. Script letters $\mathscr{R}(.)$ and $\mathscr{I}(.)$ are used to denote real and imaginary parts respectively.

One special symbol ⊰ is introduced in §84, but otherwise standard symbols are used. The notation $(\cdot, \cdot)$ is used to denote a scalar product; $\langle \cdot, \cdot \rangle$ is used generally to denote an ordered pair, but on occasion it is also used to denote a quasi-scalar-product (§7). Its meaning will, however, be clear from the context. Braces, $\{\ldots\}$, are used to indicate sets and sequences, but again the meaning will be clear from the context.

7

The spectrum and perturbations of self-adjoint operators

§93. The continuous spectrum of a self-adjoint operator

In §48 we explained how the points of the spectrum of a self-adjoint operator are classified, and then in §82 we established that the complete spectrum $\mathscr{S}(A)$ of a self-adjoint operator A coincides with the set of points of growth of the spectral function E_t and that the set of points of discontinuity of the function E_t coincides with the set of all eigenvalues of the operator A, *i.e.*, with its point spectrum $\mathscr{D}(A)$.

In order to characterize the continuous spectrum $\mathscr{C}(A)$ of a self-adjoint operator A in terms of the spectral function, we shall adopt a new definition of $\mathscr{C}(A)$, and later (in Theorem 3 of this section) we shall show that the new definition is equivalent to the earlier one in §48.

Definition. The *continuous spectrum* $\mathscr{C}(A)$ of a self-adjoint operator A (it is also called the *limit spectrum* or the *condensation spectrum*) is the set of all non-isolated points of growth of the resolution of the identity belonging to the operator A, and of all eigenvalues of infinite multiplicity.

We note immediately that this definition implies that the set $\mathscr{C}(A)$ is closed.

The points of the continuous spectrum, like the eigenvalues of the operator A, can be described by means of the equation

$$Af - \lambda f = 0. \tag{1}$$

However, in contrast to the points $\lambda \in \mathscr{D}(A)$, for which equation (1) has exact non-trivial solutions, the points $\lambda \in \mathscr{C}(A)$ are characterized by the approximate non-trivial solubility of the equation (1) with these λ. The precise meaning of this statement is expressed by the following

Theorem 1. *A point λ belongs to the continuous spectrum $\mathscr{C}(A)$ of an operator A if and only if there is in* D(A) *an infinite, orthonormal sequence*

of elements f_n such that.

$$\lim_{n\to\infty} (Af_n - \lambda f_n) = 0\,. \tag{2}$$

Proof. If $\lambda \in \mathscr{C}(A)$, then λ is either an eigenvalue of infinite multiplicity or is a non-isolated point of growth of the function E_t. In the first case, we may take as $\{f_n\}_1^\infty$ any infinite, orthonormal sequence of elements from the eigenspace corresponding to λ. In the second case, for any $\delta>0$ there is a positive $\delta'<\delta$ such that the sub-space $(E_{\lambda+\delta'}-E_{\lambda-\delta'})$H is a proper sub-set of the sub-space $(E_{\lambda+\delta}-E_{\lambda-\delta})$H. Therefore a normalized vector f can be found in $(E_{\lambda+\delta}-E_{\lambda-\delta})$H which is orthogonal to $(E_{\lambda+\delta'}-E_{\lambda-\delta'})$H. We pick out a sequence of positive numbers δ_n, decreasing without bound, such that for each δ_n the space $(E_{\lambda+\delta_n}-E_{\lambda-\delta_n})$H is a proper sub-set of the space $(E_{\lambda+\delta_{n-1}}-E_{\lambda-\delta_{n-1}})$H, and we construct an infinite sequence of elements $f_n \in (E_{\lambda+\delta_n}-E_{\lambda-\delta_n})$H for which

$$\|f_n\| = 1 \qquad \text{and} \qquad f_n \perp (E_{\lambda+\delta_{n-1}} - E_{\lambda-\delta_{n-1}})\mathrm{H}.$$

The vectors f_n satisfy the relation (2), since

$$\|Af_n - \lambda f_n\|^2 = \int_{\lambda-\delta_n}^{\lambda+\delta_n} (t-\lambda)^2\, d(E_t f_n, f_n) \leqslant \delta_n^2.$$

This proves one half of the theorem.

To prove the converse assertion, we assume that there is an orthonormal sequence of vectors $\{f_n\}_1^\infty$ satisfying the relation (2), and that, contrary to the theorem, $\lambda \notin \mathscr{C}(A)$. This means that either λ is a point of constancy of the spectral function E_t or it is an eigenvalue of finite multiplicity of the operator A which is isolated from other points of growth of the function E_t. In the first case, we shall have for some $\delta>0$ the relation

$$(E_{\lambda+\delta} - E_{\lambda-\delta})\mathrm{H} = 0\,,$$

which yields the inequality

$$\|Af - \lambda f\|^2 = \int_{-\infty}^{\infty} (t-\lambda)^2\, d(E_t f, f) \geqslant \delta^2\, \|f\|^2 \tag{3}$$

for any $f \in \mathrm{D}(A)$. This inequality is incompatible with the existence, assumed above, of the sequence $\{f_n\}_1^\infty$. In the second case, let G(λ) denote the eigenspace of the operator A corresponding to the eigenvalue λ. This sub-space is finite-dimensional; let $g_1, g_2, \ldots, g_r$ be any orthonormal basis in it. We adjoin these r vectors to the sequence $\{f_n\}_1^\infty$ and orthonormalize the resulting sequence so that the first r vectors are the $g_1, g_2, \ldots, g_r$ and we denote the subsequent vectors by $g_{r+1}, g_{r+2}, \ldots$.

The relation (2) can now be rewritten as

$$\lim_{n\to\infty} (Ag_n - \lambda g_n) = 0. \tag{2'}$$

On the other hand, for some $\delta > 0$ and for any element $f \in D(A)$ which is orthogonal to the sub-space $G(\lambda)$ we shall have the inequality

$$\|Af - \lambda f\| \geqslant \delta \, \|f\|. \tag{3'}$$

Therefore, for any $n > r$,

$$\|Ag_n - \lambda g_n\| \geqslant \delta. \tag{3''}$$

Since the relations (2′) and (3″) are incompatible, the proof of Theorem 1 is completed.

The following two theorems supplement the results of §48 and §49. As in §49, A' denotes the restriction of the operator A to the invariant sub-space $H' = H \ominus G(\lambda)$ if $\lambda \in \mathscr{D}(A)$ and $G(\lambda)$ is the eigenspace corresponding to λ, and $A' = A$ if $\lambda \notin \mathscr{D}(A)$.

Theorem 2. 1°. *If $\lambda \notin \mathscr{C}(A)$, then the operator $(A' - \lambda I)^{-1}$ is bounded.*
2°. *If $\lambda \in \mathscr{C}(A)$, and if λ is not an eigenvalue of infinite multiplicity of the operator A, then the operator $(A' - \lambda I)^{-1}$ is unbounded.*

Proof. If $\lambda \notin \mathscr{C}(A)$, then, for all $f \in D(A')$ and for some $\delta > 0$, the inequality (3′) holds, *i.e.*,

$$\|A'f - \lambda f\| \geqslant \delta \, \|f\|,$$

which implies that

$$\|(A' - \lambda I)^{-1}\| \leqslant 1/\delta,$$

and assertion 1° is proved.

Suppose the hypotheses of 2° hold. Then λ is a non-isolated point of growth of the resolution of the identity E_t' for the operator A' in H′, *i.e.*, $\lambda \in \mathscr{C}(A')$ but $\lambda \notin \mathscr{D}(A')$. By Theorem 1 there is an infinite, orthornormal sequence of vectors $f_n \in H'$ ($n \in \mathbf{N}$) for which

$$\lim_{n\to\infty} (A'f_n - \lambda f_n) = 0,$$

and this implies that the operator $(A' - \lambda I)^{-1}$ is unbounded. Q.E.D.

The next theorem shows, in particular (see 3° and 4°), that the definition of the continuous spectrum adopted in this section is equivalent to the definition in §48.

Theorem 3.

1°. *The relation* $\lambda \notin \mathcal{S}(A)$ *is equivalent to the equality* $R(A-\lambda I)=H$.

2°. *The relation* $\lambda \in \mathcal{D}(A)$ *is equivalent to the inequality* $\overline{R(A-\lambda I)} \neq H$.

3°. *If* $\lambda \notin \mathcal{C}(A)$, *then* $\overline{R(A-\lambda I)}=R(A-\lambda I)$.

4°. *If* $\lambda \in \mathcal{C}(A)$ *and if* λ *is not an eigenvalue of infinite multiplicity, then* $\overline{R(A-\lambda I)} \neq R(A-\lambda I)$.

Proof. Assertions 1° and 2° hold by virtue of Definition 2 and Theorem 2 of §48.

If $\lambda \notin \mathcal{C}(A)$, then, by Theorem 2 of this section, the operator $(A-\lambda I)^{-1}$ is bounded. Hence and because the operator A' is closed, it follows that the manifold $R(A'-\lambda I)$ is closed. And since, in accordance with §49, $R(A'-\lambda I)=\overline{R(A-\lambda I)}$, assertion 3° holds.

Finally, if the hypothesis of 4° holds, then, again by Theorem 2, the symmetric operator $(A'-\lambda I)^{-1}$ is unbounded. Therefore (see §25) its domain $R(A'-\lambda I)=\overline{R(A-\lambda I)}$ cannot coincide with the whole sub-space $\overline{R(A-\lambda I)}$. Q.E.D.

A criterion that there shall be no points of the continuous spectrum of a self-adjoint operator A in a given interval is provided by

Theorem 4. If the closed interval $[\lambda_0-\rho, \lambda_0+\rho]$ *does not contain points of* $\mathcal{C}(A)$, *then there is a finite-dimensional sub-space* $G \subset H$ *such that, for any vector* $f \neq 0$ *in* $D(A) \cap (H \ominus G)$, *the inequality*

$$\|(A-\lambda_0 I)f\| > \rho \, \|f\| \tag{4}$$

holds.

On the other hand, if there is a finite-dimensional sub-space $G \subset H$ *such that, for all* $f \in D(A) \cap (H \ominus G)$, *the inequality*

$$\|(A-\lambda_0 I)f\| \geqslant \rho \, \|f\| \tag{4'}$$

holds, then the open interval $(\lambda_0-\rho, \lambda_0+\rho)$ *contains no points of* $\mathcal{C}(A)$.

Proof. If the set $\mathcal{C}(A)$ has no points in the interval $[\lambda_0-\rho, \lambda_0+\rho]$, then the part of $\mathcal{S}(A)$ lying in this interval consists entirely of a finite number of eigenvalues of finite multiplicity. Let G denote the linear envelope of all the eigenvectors corresponding to these eigenvalues. Then, for some $\delta > \rho$, we have the inequality

$$\|(A-\lambda_0 I)f\|^2 = \int_{-\infty}^{\infty} (t-\lambda_0)^2 \, d(E_t f, f) \geqslant \delta^2 \, \|f\|^2$$

provided only that $f \in D(A)$ and $f \perp G$. This gives (4) (when $f \neq 0$).

Now suppose that the condition (4′) is satisfied, but that, contrary to the theorem, the set $(\lambda_0-\rho, \lambda_0+\rho)\cap\mathscr{C}(A)$ is not empty. This implies that the sub-space $(E_{\lambda_0+\rho}-E_{\lambda_0-\rho+0})$H is infinite-dimensional. Therefore there is in this sub-space a vector $f\neq 0$ which is orthogonal to $(E_{\lambda_0+\rho}-E_{\lambda_0-\rho+0})$G. This vector f clearly belongs to D(A) and is orthogonal to G. Therefore the inequality (4′) must hold for this f. On the other hand, for this f,

$$\|(A-\lambda_0 I)f\|^2=\int_{\lambda_0-\rho+0}^{\lambda_0+\rho}(t-\lambda_0)^2\,d(E_tf,f)$$

$$<\rho^2\int_{\lambda_0-\rho+0}^{\lambda_0+\rho}d(E_tf,f)=\rho^2(f,f).$$

We have a contradiction; hence the theorem.

Finally, it should be noted that for an arbitrary closed, but not self-adjoint, operator T the point spectrum $\mathscr{D}(T)$ is defined as the set of eigenvalues, and the continuous spectrum as the set $\mathscr{C}(T)$ of those λ for which there exists a bounded, non-compact sequence of vectors $\{f_n\}_1^\infty$ satisfying the condition (2), *viz.*,

$$\lim_{n\to\infty}(Af_n-\lambda f_n)=0.$$

In this case the spectrum does not, in general, contain just the points of the sets $\mathscr{D}(T)$ and $\mathscr{C}(T)$, but may also contain what is called the *residual spectrum*, namely, those values of λ for which R$(T-\lambda I)$ is not dense in H and which are not eigenvalues of the operator T.

§94. H. Weyl's theorem and von Neumann's theorem on completely continuous perturbations

The following remarkable theorem is due to H. Weyl (see Weyl [126]).

Theorem 1. If a completely continuous, self-adjoint operator K is added to a self-adjoint operator A, then the continuous spectrum of the operator A remains unchanged, i.e.,

$$\mathscr{C}(A+K)=\mathscr{C}(A).$$

Proof. From the spectral resolution of completely continuous, self-adjoint operators it follows that, for any $\varepsilon>0$, the operator K can be expressed in the form

$$K=F+B,$$

where F is a finite-dimensional, self-adjoint operator,
and B is a self-adjoint operator such that $\|B\| \leq \varepsilon$.
Therefore Theorem 1 can be proved by proving the following simpler propositions.

Lemma 1. If A is an arbitrary self-adjoint operator, and F a finite-dimensional self-adjoint operator, then

$$\mathscr{C}(A+F) = \mathscr{C}(A).$$

Lemma 2. If A is an arbitrary self-adjoint operator and if B is a bounded, self-adjoint operator, then the continuous spectrum $\mathscr{C}(A+B)$ of the perturbed operator lies in a closed $\|B\|$-neighbourhood of the set $\mathscr{C}(A)$.

Both these are proved by means of Theorem 4 of §93.

Proof of Lemma 1. We introduce the sub-space $G = R(F)$. Since the operator F is equal to 0 on $H \ominus G$, we have

$$\|(A+F)f - \lambda f\| = \|Af - \lambda f\|$$

for any λ and for any $f \in D(A) \cap (H \ominus G)$. Hence, by Theorem 4 of §93 —— which is applicable here because $R(F)$ is finite-dimensional, we conclude that if a closed interval $[\alpha, \beta]$ does not contain points of *one* of the sets $\mathscr{C}(A)$ and $\mathscr{C}(A+F)$, then the corresponding open interval (α, β) also does not contain points of the other one of these two sets. And since both of the sets $\mathscr{C}(A)$ and $\mathscr{C}(A+F)$ are closed, they coincide.

Proof of Lemma 2. We observe first of all that in the statement of this lemma, the operators A and $A+B$ could be interchanged. It therefore suffices to prove that a closed $\|B\|$-neighbourhood of each point of the set $\mathscr{C}(A)$ contains at least one point of the set $\mathscr{C}(A+B)$.

To do this we assume the contrary, and suppose that in a closed $\|B\|$-neighbourhood of some point $\lambda \in \mathscr{C}(A)$ there are no points of $\mathscr{C}(A+B)$. It then follows from Theorem 4 of §93 that there is a sub-space G and a $\rho > \|B\|$ such that, for any non-zero element $f \in D(A)$ which is orthogonal to G, *i.e.*, $f \in D(A) \cap (H \ominus G)$, the inequality

$$\|(A+B-\lambda I)f\| > \rho \, \|f\|$$

holds. But then, for the same function f, we have

$$\|(A-\lambda I)f\| \geq \|(A+B-\lambda I)f\| - \|Bf\| > (\rho - \|B\|) \, \|f\|,$$

and therefore, by Theorem 4 of §93 again, it follows that $\lambda \notin \mathscr{C}(A)$, contrary to hypothesis. Hence Lemma 2, and with it also Theorem 1, has been proved.

We remark that Weyl's theorem can be generalized to the case of an arbitrary linear, closed operator A and any completely continuous operator K, *i.e.*, the requirement that these operators be self-adjoint can be dropped (see, *e.g.*, Glazman [37]).

While Theorem 1 shows that a completely continuous perturbation K of an operator A does not affect its continuous spectrum, the theorem says nothing about the effect of K on the point spectrum $\mathcal{D}(A)$. The following theorem, due to von Neumann, shows how great the effect on $\mathcal{D}(A)$ may be.

Theorem 2. To any self-adjoint operator A operating in a separable Hilbert space H *can be added a self-adjoint operator K which is not only completely continuous but may even have an arbitrarily small norm, and which is such that the set of all eigenvalues of the operator $A+K$ will be complete in* H.

(Thus the set $\mathcal{D}(A+K)$ will be dense in $\mathcal{C}(A)=\mathcal{C}(A+K)$ even though the set $\mathcal{D}(A)$ may be empty.)

Proof. It suffices to consider the case where the operator A is bounded. For, an unbounded operator A can be expressed as a sum of bounded, self-adjoint operators

$$A_j = E(\Delta_j)A, \quad \text{where} \quad \Delta_j = [j-1, j) \qquad (-\infty < j < \infty).$$

If the theorem is proved for the operators A_j, then we can construct completely continuous operators K_j which operate in the orthogonal sub-spaces $E(\Delta_j)$H and are such that the absolute norms

$N(K_j) < \delta/(j^2+1)$, where δ is an arbitrary positive number.
But then the operator

$$K = \sum_{j=-\infty}^{\infty} \oplus K_j$$

will be the required perturbation of the operator K. In fact, thanks to the triangle inequality for absolute norms (see §31)

$$N(K) \leq \sum_{j=-\infty}^{\infty} N(K_j) < 4\delta.$$

We assume, then, that the operator A is bounded.

Von Neumann's proof (see von Neumann [95])† given below is based

† A theorem similar to Theorem 2 was proved earlier by Weyl but without the assertion of the possibility of restriction to completely continuous perturbations of Hilbert–Schmidt type.

on the following lemma, which, for the moment, we merely formulate; we shall prove it later.

Lemma 3. *If A is a bounded, self-adjoint operator and if* $g \neq 0$ *is any element in* H, *then, for any* $\delta > 0$, *there is a finite-dimensional sub-space* $G \subset H$ *and a finite-dimensional, self-adjoint operator F such that:*

1°. $g \in G$;
2°. G reduces $A + F$;
3°. $N(F) \leqslant \delta$.

By means of this lemma we can construct, in a countable number of steps, a completely continuous, self-adjoint operator K and a complete, orthogonal system of finite-dimensional sub-spaces which are invariant under the operator $A + K$. Clearly, this operator $A + K$ will have a set of eigenvalues which is dense in H.

To carry out this construction, we select a sequence $\{f_k\}_1^\infty$ everywhere dense in H, and a number $\varepsilon > 0$. Applying Lemma 3 to the element $g = g_1 = f_1$ and the number $\delta = \varepsilon/2$, we find a finite-dimensional sub-space G_1 and a finite-dimensional, self-adjoint operator K_1 with absolute norm $N(K_1) \leqslant \varepsilon/2$; the operator $A + K_1$ is reduced by the sub-space G_1 and therefore also by the space $H \ominus G_1$.

Since the sequence $\{f_k\}_1^\infty$ is dense in H, not all the elements f_k can be in G_1. Without loss of generality, we can assume that f_2 is not in G_1. In that case, we let g_2 be the projection of f_2 on to $H \ominus G_1$, and we apply Lemma 3 to $g = g_2$ and the number $\delta = \varepsilon/2^2$, with H replaced by $H \ominus G_1$ and A replaced by $A + K_1$. We shall thus obtain a finite-dimensional sub-space $G_2 \subset H \ominus G_1$ and an operator K_2 in $H \ominus G_1$. Moreover, the operator $A + K_1 + K_2$ in $H \ominus G_1$ is reduced by the sub-space G_2 and $N(K_2) \leqslant \varepsilon/2^2$. If we extend K_2 linearly on to the whole space H by putting $K_2 = 0$ in G_1, then clearly the inequality $N(K_2) \leqslant \varepsilon/2^2$ will still hold, and the operator $A + K_1 + K_2$ will still be reduced by the sub-space G_2.

Continuing this process, we arrive at an infinite sequence of mutually orthogonal sub-spaces G_j and a sequence of finite-dimensional operators K_j with absolute norms $N(K_j) \leqslant \varepsilon/2^j$ and such that, for each j, the operator $A + K_1 + K_2 + \cdots + K_j$ is reduced by each of the sub-spaces $G_1, G_2, \ldots, G_j$.

We now define an operator K by putting

$$K = \sum_{j=1}^{\infty} K_j.$$

This operator series converges uniformly, and the inequality $N(K) \leqslant \varepsilon$ is obtained from the above.

Clearly, the operator $A+K$ is reduced by each of the sub-spaces G_j $(j=1, 2, 3, \ldots)$, and it remains to show that the sub-space

$$G = G_1 \oplus G_2 \oplus \cdots \oplus G_j \oplus \cdots$$

coincides with H. To do this, it suffices to show that each of the vectors f_k $(k=1, 2, \ldots)$ is orthogonal to $H \ominus G$. But, by construction, $f_1 = g_1 \in G_1$, and so $f_1 \perp H \ominus G_1$ and therefore, *a fortiori*, $f_1 \perp H \ominus G$. Further, $f_2 = g_2 + h_2$, where $g_2 \in G_2$ and $h_2 \perp H \ominus G_1$, from which it follows that $f_2 \perp H \ominus (G_1 \oplus G_2)$, and therefore, *a fortiori*, $f_2 \perp H \ominus G$. These considerations show that, for any natural number n, the relation

$$f_n \perp H \ominus \sum_{j=1}^{n} \oplus G_j$$

holds, from which it follows that $f_n \perp H \ominus G$.

This proves Theorem 2 except that we still have to prove Lemma 3.

To do this, let α, β be the lower and upper bounds of the operator A. We divide the interval $[\alpha, \beta]$ into n equal parts by points γ_k $(k=0, 1, 2, \ldots, n)$. The method of choosing the integer $n = n(\delta)$ will be given later.

Writing

$$\Delta_k = [\gamma_{k-1}, \gamma_k] \qquad (k=1, 2, \ldots, n-1; \gamma_0 = \alpha)$$

and

$$\Delta_n = [\gamma_{n-1}, \gamma_n] \qquad (\gamma_n = \beta),$$

we form the n vectors $h_k = E(\Delta_k)g$ $(k=1, 2, \ldots, n)$ and denote their linear envelope by G, and the operator of orthogonal projection on to G by P.

By normalizing the vectors h_k we obtain an orthonormal basis $\{g_k\}_1^\infty$ for the sub-space G. If some of the vectors h_k are null, then the dimension of G will be less than n, but this will not affect the argument, and we may for convenience take $\dim G = n$.

Further, we put

$$F_1 = -(I-P)AP$$

and we define a finite-dimensional operator F by

$$F = F_1 + F_1^*.$$

Then

$$A + F = (I-P)A(I-P) + PAP,$$

and so the operator $A+F$ commutes with P, and therefore it is reduced by the sub-space G.

It remains to estimate the absolute norm $N(F)$. By completing the sequence $\{g_k\}_1^n$ to an orthonormal basis $\{g_k\}_1^\infty$ in H, we shall have

$$N^2(F_1)=\sum_{j,k=1}^{\infty}|(F_1g_j,\,g_k)|^2=\sum_{\substack{1\leq j\leq n\\1\leq k<\infty}}|(F_1g_j,\,g_k)|^2=\sum_{j=1}^{n}\|F_1g_j\|^2. \tag{1}$$

Since $g_j\in E(\Delta_j)\mathrm{H}$, we have

$$\|Ag_j-\gamma_jg_j\|\leq(\beta-\alpha)/n,$$

or

$$Ag_j=\gamma_jg_j+f_j,\qquad \text{where}\quad \|f_j\|\leq(\beta-\alpha)/n.$$

Replacing g_j in this equality by $Pg_j=g_j$ and applying the operator $(P-I)$ to both sides, we obtain the relation

$$F_1g_j=(P-I)APg_j=(P-I)f_j$$

and the inequality

$$\|F_1g_j\|\leq\|f_j\|\leq(\beta-\alpha)/n.$$

The relation (1) now gives

$$N(F_1)\leq(\beta-\alpha)/\sqrt{n}$$

and so

$$N(F)\leq N(F_1)+N(F_1^*)=2N(F_1)\leq2(\beta-\alpha)/\sqrt{n}.$$

By choosing n sufficiently large, we obtain the required inequality $N(F)\leq\delta$, and thus Lemma 3 is proved.

In connection with Theorem 1 the following question arises: what can be said regarding the 'closeness' of two bounded, self-adjoint operators A and B if their continuous spectra coincide? It is easy to see from examples that the difference $A-B$ of two such operators may fail to be a completely continuous operator. So a direct converse of Theorem 1 is not possible. However, the following theorem (see von Neumann [95]) holds.

Theorem 3. *If the space* H *is separable, and if A and B are bounded, self-adjoint operators in* H *and have the same continuous spectrum, then there is a unitary operator U such that the operator*

$$K=B-UAU^{-1} \tag{2}$$

will be completely continuous.

Proof. By Theorem 2 we may suppose that each of the operators A and B has an orthonormal system of eigenvectors complete in H; denote these by $\{e_k\}_1^\infty$ for A and $\{f_k\}_1^\infty$ for B, and let $\{\lambda_k\}_1^\infty$ and $\{\mu_k\}_1^\infty$ be the corresponding sequences of eigenvalues. If, for some permutation $\{p_k\}_1^\infty$ of the natural numbers,

$$\lim_{k\to\infty} (\lambda_k - \mu_{p_k}) = 0, \tag{3}$$

then, defining a unitary operator U by the relations

$$Ue_p = f_{p_k} \qquad (k = 1, 2, \ldots),$$

and a self-adjoint operator K by (2), we find that

$$Kf_{p_k} = (\mu_{p_k} - \lambda_k) f_{p_k} \qquad (k = 1, 2, \ldots),$$

and this implies, because of (3), that the operator K is completely continuous.

Thus the proof of the theorem is reduced to the construction of the permutation $\{p_k\}$ for which the relation (3) holds. The possibility of such a construction depends on the coincidence of the set $\mathscr{C}(A)$ of limit points of the sequence $\{\lambda_k\}_1^\infty$ with the set $\mathscr{C}(B)$ of limit points of the sequence $\{\mu_k\}_1^\infty$. The relevant proposition in the theory of limits of numerical sequences may be formulated as follows.

If the set M *of limit points of a bounded sequence of real numbers* $\{\lambda_k\}_1^\infty$ *coincides with the limit points of a similar sequence* $\{\mu_k\}_1^\infty$, *then there is a permutation* $\{p_k\}_1^\infty$ *of the natural numbers such that*

$$\lim_{k\to\infty} (\lambda_k - \mu_{p_k}) = 0.$$

To prove this (see von Neumann [95]) we define, for any natural number k, the numbers

$$\varepsilon_k = \min_{t\in M} |\lambda_k - t| + \frac{1}{k}, \qquad \eta_k = \min_{t\in M} |\mu_k - t| + \frac{1}{k}.$$

Since, for any $\delta > 0$, only a finite number of the points λ_k, μ_k lie at a distance $>\delta$ from M, it follows that

$$\lim_{k\to\infty} \varepsilon_k = \lim_{k\to\infty} \eta_k = 0.$$

By definition of the numbers ε_k, it follows that, for any k, the interval $(\lambda_k - \varepsilon_k, \lambda_k + \varepsilon_k)$ contains points of M, and so this interval contains an infinite number of the points μ_k. Let r_k be the least subscript r such that $\mu_r \in (\lambda_k - \varepsilon_k, \lambda_k + \varepsilon_k)$ and which satisfies the inequalities $r > r_j$ $(j = 1, 2, \ldots, k-1)$, $r > 2k$. Similarly, starting from the interval

$(\mu_k - \eta_k, \mu_k + \eta_k)$, we define an integer s_k. All the subscripts r_k are distinct, and so are the s_k, and

$$\lim_{k\to\infty} (\lambda_k - \mu_{r_k}) = \lim_{k\to\infty} (\mu_k - \lambda_{s_k}) = 0. \tag{4}$$

If either one or both of the sequences $\{r_k\}_1^\infty$, $\{s_k\}_1^\infty$ contained all the natural numbers, the theorem would be proved. However, here this condition is not satisfied. We therefore continue the construction by defining inductively two sequences $\{u_k\}_1^\infty$ and $\{v_k\}_1^\infty$ of natural numbers by putting

$$u_1 = 1, \qquad v_1 = r_1$$

and letting

v_{2k} be the least natural number distinct from v_j $\quad (j = 1, 2, \ldots, 2k-1)$,

$u_{2k} = s_{v_{2k}}$,

u_{2k+1} be the least natural number distinct from u_j $\quad (j = 1, 2, \ldots, 2k)$,

$v_k = r_{u_{2k+1}}$.

We shall prove that each of the sequences $\{u_k\}_1^\infty$, $\{v_k\}_1^\infty$ so constructed is a permutation of the natural numbers. We do this by showing that each of these sequences contains all the natural numbers and contains each natural number once only. If the integers $1, 2, \ldots, k-1$ are contained in the sequence $\{u_j\}_1^{2k-2}$, then either the integer k is also contained in this sequence, or $u_{2k-1} = k$. In either case the sequence $\{u_j\}_1^{2k}$ contains the integers $1, 2, \ldots, k$. Since the sequence $\{u_j\}_1^2$ contains the integer 1, it follows by induction that the sequence $\{u_j\}_1^\infty$ contains all the natural numbers. The sequence $\{v_j\}_1^\infty$ also has the same property.

It remains to show that neither of the sequences $\{u_j\}_1^\infty$, $\{v_j\}_1^\infty$ contains two equal integers.

We start with $\{u_j\}_1^\infty$. Since by construction, $u_{2k-1} \neq u_1, u_2, \ldots, u_{2k-2}$, we have only to show that $u_{2k} \neq u_1, u_2, \ldots, u_{2k-1}$. By definition, $u_{2k} = s_{v_{2k}}$, and $s_{v_{2k}} \neq s_1, s_2, \ldots, s_{v_{2k-1}}$. Therefore, in particular, $s_{v_{2k}} \neq s_{v_2}, s_{v_4}, \ldots, s_{v_{2k-2}}$, because $v_2 < v_4 < \cdots < v_{2k-2}$ and the sequence $\{s_k\}_1^\infty$ is monotone increasing. The inequality obtained can be written as $u_{2k} \neq u_2, u_4, \ldots, u_{2k-2}$, and it remains to show that $u_{2k} \neq u_1, u_3, \ldots, u_{2k-1}$. But, by construction, $u_{2k} = s_{v_{2k}} \geq 2v_{2k} \geq 2k$, and $u_1, u_3, \ldots, u_{2k-1} \leq 2k-1$, so that $u_{2k} > u_{2k-1}$, and therefore $u_{2k} \neq u_1, u_3, \ldots, u_{2k-1}$.

Turning now to the sequence $\{v_k\}_1^\infty$, we note first that, by construction, $v_{2k-1} = r_{u_{2k-1}} \neq r_1, r_2, \ldots, r_{u_{2k-3}}$, and so, in particular, $v_{2k-1} \neq v_1, v_3, \ldots, v_{2k-3}$. Moreover, $v_{2k-1} = r_{u_{2k-1}} \geq 2u_{2k-1} \geq 2k$, and since $v_2, v_4, \ldots, v_{2k-2} \leq 2k-2$, we have $v_{2k-1} \neq v_1, v_2, \ldots, v_{2k-2}$. Finally, by construction, $u_{2k} \neq v_1, v_2, \ldots, v_{2k-1}$.

To complete the proof, it is enough to verify that

$$\lim_{k\to\infty} (\mu_{v_k} - \lambda_{u_k}) = 0.$$

But this follows from the fact that, as $k\to\infty$,

$$|\mu_{v_{2k}} - \lambda_{u_{2k}}| = |\mu_{v_{2k}} - \lambda_{s_{v_{2k}}}| \to 0,$$
$$|\mu_{v_{2k-1}} - \lambda_{u_{2k-1}}| = |\mu_{r_{u_{2k-1}}} - \lambda_{u_{2k-1}}| \to 0, \text{ by (4)}.$$

§95. The absolutely continuous part and the singular part of the spectrum

In connection with the so-called 'dispersion operators' or 'wave operators' which occur in mathematical physics (see also §97), it has in recent years become necessary to have a more precise classification of the points of the spectrum than that which was adopted in §82 and §93.

Let A be a self-adjoint operator and E_t its spectral function.

An element $f\in \mathrm{H}$ is said to be *regular* relative to A if the function $\sigma(t;f)=(E_t f, f)$ is locally absolutely continuous, and *singular* if the absolutely continuous part of the function $\sigma(t;f)$ is equal to 0.

Let H_a denote the set of all elements which are regular relative to A, and H_s the set of all singular elements.

Theorem 1. The sets H_a *and* H_s *are sub-spaces of* H; *they are orthogonal, and*

$$\mathrm{H}_a \oplus \mathrm{H}_s = \mathrm{H}.$$

Proof. Let $f\in \mathrm{H}_a$, $g\in \mathrm{H}_s$. Then, for any point-set e of Lebesgue measure zero,

$$\int_e d(E_t f, f) = 0, \tag{1}$$

and there is a Borel set† θ of Lebesgue measure zero such that, on its complement $C\theta$,

$$\int_{C\theta} d(E_t g, g) = 0. \tag{2}$$

We now use the inequality

$$\left|\int_E d(E_t f, g)\right|^2 \leqslant \int_E d(E_t f, f) \int_E d(E_t g, g),$$

† See, e.g., Smirnov [118] or Shilov and Gurevich [113].

which holds for any Borel set $E\subset(-\infty,\infty)$; this is the Cauchy–Bunyakovski inequality for the quasi-scalar product (see §3)

$$\langle f, g\rangle = \int_E d(E_t f, g).$$

It follows from (1) and (2), by virtue of this inequality, that

$$\int_\theta d(E_t f, g) = 0, \qquad \int_{C\theta} d(E_t f, g) = 0.$$

Therefore

$$(f, g) = \int_{-\infty}^{\infty} d(E_t f, g) = 0,$$

which implies that $H_a \perp H_s$.

We now select an arbitrary non-zero vector $h\in H$; to it corresponds a non-decreasing function $\sigma(t; h) = (E_t h, h)$, from which we pick out the absolutely continuous part $\sigma_a(t; h)$:

$$\sigma(t; h) = \sigma_a(t; h) + \sigma_s(t; h).$$

Then we choose a Borel set θ of Lebesgue measure zero whose complement $C\theta$ has zero σ_s-measure, and we put

$$f = \int_{C\theta} dE_u h, \qquad g = \int_\theta dE_u h, \tag{3}$$

so that

$$h = f + g. \tag{4}$$

Then (3) implies

$$(E_t f, f) = \int_{-\infty}^{t} \chi_{C\theta}(u)\, d(E_u h, h) = \int_{-\infty}^{t} d\sigma_a(u; h),$$

$$(E_t g, g) = \int_{-\infty}^{t} \chi_\theta(u)\, d(E_u h, h) = \int_{-\infty}^{t} d\sigma_s(u; h),$$

where $\chi_{C\theta}$ and χ_θ are the characteristic functions of the sets $C\theta$ and θ. It follows from these representations that the function $(E_t f, f)$ is absolutely continuous, and that the absolutely continuous part of the function $(E_t g, g)$ is null. Therefore the elements f and g in the representation (4) belong to H_a and H_s respectively.

From the resolution of an arbitrary element $h\in H$ into the orthogonal sum (4) of an element $f\in H_a$ and an element $g\in H_s$ it follows that the sets H_a and H_s are linear and closed. This concludes the proof of the theorem.

Theorem 2. The sub-spaces H_a *and* H_s *reduce the operator A.*

Proof. It suffices to show that, for an arbitrary λ, the sub-space H_a reduces E_λ. If $f \in H_a$, then the function $\sigma(t; f)$ is absolutely continuous; but then the function

$$\sigma(t; E_\lambda f) = (E_t E_\lambda f, E_\lambda f) = (E_{\min(t,\lambda)} f, f)$$
$$= \begin{cases} \sigma(t; f) & (t \leq \lambda) \\ \sigma(\lambda; f) = \text{const.} & (t > \lambda) \end{cases}$$

is also absolutely continuous, and so $E_\lambda f \in H_a$.

Similarly it can be proved that if $f \in H_s$, then $E_\lambda f \in H_s$. Q.E.D.

The restriction A_a of the operator A to the sub-space H_a of regular elements is called the *absolutely continuous part of the operator* A, and the spectrum $\mathscr{S}(A_a)$ of this operator A_a is called *the absolutely continuous part of the spectrum of the operator* A.

It is clear that $\mathscr{S}(A_a)$ forms a closed set without isolated points (*i.e.*, a perfect set), and $\mathscr{S}(A_a) \subseteq \mathscr{C}(A)$.

The restriction A_s of the operator A to the sub-space H_s of singular elements is called the *singular part of the operator* A, and its spectrum $\mathscr{S}(A_s)$ is the *singular part of the spectrum of the operator* A. Clearly, $\mathscr{D}(A) \subseteq \mathscr{S}(A_s)$.

The eigenvalues —— all of which come into $\mathscr{S}(A_s)$ —— form the *discrete component* of the singular part of the spectrum of the operator A. The complement of this discrete component in the whole of $\mathscr{S}(A_s)$ is called the *continuous component* of the spectrum of the operator A.

§96. Invariance of the absolutely continuous part of the spectrum relative to finite-dimensional perturbations

It follows from Theorem 2 of §94 that, with absolutely continuous perturbations of a self-adjoint operator, the absolutely continuous part of the spectrum may not be preserved——in contrast to the behaviour of the continuous part. But under weaker perturbations, and, in particular, under finite-dimensional perturbations, the absolutely continuous part of the spectrum does not change. This result, and its generalization to the case of arbitrary nuclear perturbations, is due to M. Rozenblum and to T. Kato (see Rozenblum [111] and Kato [47, 48]).

Let A be an arbitrary self-adjoint operator in a Hilbert space H; we do not assume that H is separable. Let K be a 1-dimensional operator generated by some unit vector g and some real number γ in the sense that, for any $f \in \mathrm{H}$,

$$Kf = \gamma(f, g)g. \tag{1}$$

Let E_t be the spectral function of the operator A, and F_t that of the operator $B = A + K$. We make $[\alpha, \beta)$ traverse the set of all left-closed intervals of the real line. Corresponding to the set of such intervals there is a set of elements $(E_\beta - E_\alpha)g$, whose closed envelope we denote by $\mathrm{G_A}$, and also a set of elements $(F_\beta - F_\alpha)g$, whose closed linear envelope we denote by $\mathrm{G_B}$. It is clear that $\mathrm{G_A}$ is the smallest sub-space of H which contains g and which reduces the operator A; and $\mathrm{G_B}$ has the same property as regards the operator B. Moreover, each of the sub-spaces $\mathrm{G_A}$ and $\mathrm{G_B}$ reduces the operator K. Hence it follows from the equality $B = A + K$ that $\mathrm{G_A} = \mathrm{G_B} \equiv \mathrm{H_g}$. Since $Kf = 0$ if $f \in \mathrm{H} \ominus \mathrm{H_g}$, we have

$$Af = Bf \qquad (f \in \mathrm{H} \ominus \mathrm{H_g}),$$

and the question concerning the effect of the 1-*dimensional perturbation K on the absolutely continuous part of the spectrum of the operator A need be considered only in the separable sub-space* $\mathrm{H_g}$. It is this question which we now tackle.

Since the sub-space $\mathrm{H_g}$ is closed whether it be regarded as the linear envelope of the elements $(E_\beta - E_\alpha)g$ or as the linear envelope of the elements $(F_\beta - F_\alpha)g$, it follows that the restrictions of the operators A and B to $\mathrm{H_g}$ each have a simple spectrum. Therefore (see §83), for any $f \in \mathrm{H_g}$,

$$f = \int_{-\infty}^{\infty} \varphi(t)\, dE_t g \quad \text{and} \quad f = \int_{-\infty}^{\infty} \psi(t)\, dF_t g \tag{2}$$

and

$$\|f\|^2 = \int_{-\infty}^{\infty} |\varphi(t)|^2\, d\rho(t) = \int_{-\infty}^{\infty} |\psi(t)|^2\, d\sigma(t),$$

where $\rho(t) = (E_t g, g)$ and $\sigma(t) = (F_t g, g)$.

On the other hand, if $\mathscr{I}(\zeta) \neq 0$, then

$$(B - \zeta I)^{-1} - (A - \zeta I)^{-1} = -(A - \zeta I)^{-1} K (B - \zeta I)^{-1},$$

from which, by (1), we have

$$([B - \zeta I]^{-1} f, g) - ([A - \zeta I]^{-1} f, g)$$
$$= -\gamma([B - \zeta I]^{-1} f, g)([A - \zeta I]^{-1} g, g). \tag{3}$$

Using the representations (2) we obtain from (3) the equality

$$\int_{-\infty}^{\infty} \frac{\varphi(\mu)\, d\sigma(\mu)}{\mu - \zeta} - \int_{-\infty}^{\infty} \frac{\varphi(\lambda)\, d\rho(\lambda)}{\lambda - \zeta} = -\gamma \int_{-\infty}^{\infty} \frac{\psi(\mu)\, d\sigma(\mu)}{\mu - \zeta} \int_{-\infty}^{\infty} \frac{d\rho(\lambda)}{\lambda - \zeta}$$

or

$$\int_{-\infty}^{\infty} \frac{\psi(\mu)\, d\sigma(\mu)}{\mu-\zeta}\left\{1+\gamma \int_{-\infty}^{\infty} \frac{d\rho(\lambda)}{\lambda-\zeta}\right\}=\int_{-\infty}^{\infty} \frac{\varphi(\lambda)\, d\rho(\lambda)}{\lambda-\zeta}. \tag{4}$$

If we had taken $f=g$ in (3), we would have obtained

$$\int_{-\infty}^{\infty} \frac{d\sigma(\mu)}{\mu-\zeta}\left\{1+\gamma \int_{-\infty}^{\infty} \frac{d\rho(\lambda)}{\lambda-\zeta}\right\}=\int_{-\infty}^{\infty} \frac{d\rho(\lambda)}{\lambda-\zeta},$$

which can be rewritten as

$$\left\{1+\gamma \int_{-\infty}^{\infty} \frac{d\rho(\lambda)}{\lambda-\zeta}\right\}\left\{1-\gamma \int_{-\infty}^{\infty} \frac{d\sigma(\mu)}{\mu-\zeta}\right\}=1.$$

Putting $\zeta=\xi+i\eta$, $\quad \eta=\mathscr{I}(\zeta)$, and letting $\eta\to 0$, we have, by a well-known theorem in the theory of functions (see, *e.g.*, Privalov [102])

1) almost everywhere on the real line the function

$$\omega(\zeta)\equiv 1+\gamma \int_{-\infty}^{\infty} \frac{d\rho(\lambda)}{\lambda-\zeta}=\left[1-\gamma \int_{-\infty}^{\infty} \frac{d\sigma(\mu)}{\mu-\zeta}\right]^{-1} \tag{5}$$

has finite, non-zero limits $\omega(\xi+i0)$ and $\omega(\xi-i0)=\overline{\omega(\xi+i0)}$; and

2) almost everywhere

$$\pi\gamma\rho'(\xi)=\mathscr{I}[\omega(\xi+i0)],$$

$$\pi\gamma\sigma'(\xi)=-\mathscr{I}\left[\frac{1}{\omega(\xi+i0)}\right]=\frac{\mathscr{I}[\omega(\xi+i0)]}{|\omega(\xi+i0)|^2},$$

so that, almost everywhere,

$$\rho'(\xi)=|\omega(\xi+i0)|^2\sigma'(\xi), \tag{6}$$

and therefore, almost everywhere,

$$\mathscr{I}[\omega(\xi+i0)]=\pi\gamma\,|\omega(\xi+i0)|^2\,\sigma'(\xi). \tag{6'}$$

From the relation (6) and what we have said about the function $\omega(\xi+i0)$ we conclude that the absolutely continuous parts of the spectra of the operators A and B coincide up to a set of measure zero. But it is easy to prove that, if the intersection of the absolutely continuous spectrum with some open interval is of Lebesgue measure zero, then this intersection is empty. Therefore the absolutely continuous parts of the spectra of the operators A and B coincide exactly. Moreover, the theorem of §87 implies that the absolutely continuous parts A_a and B_a of the operators A and B are unitarily invariant.

From these considerations we can formulate the following theorem.

Theorem. *Let A be an arbitrary self-adjoint operator in a Hilbert space* H, *and let K be an arbitrary finite-dimensional, self-adjoint operator in* H; *let* $B = A + K$.

Then the absolutely continuous parts A_a *and* B_a *of the operators A and B are unitarily equivalent, and so, in particular,*

$$\mathscr{S}(B_a) = \mathscr{S}(A_a).$$

Proof. The result has already been proved for the case where the operator K is 1-dimensional. Since an m-dimensional perturbation is the sum of 1-dimensional perturbations, and since the concept of unitary equivalence is transitive, the theorem is true for any finite-dimensional perturbations. Q.E.D.

The generalization, mentioned at the beginning of this section, of this theorem to the case of arbitrary, self-adjoint, nuclear perturbations must be postponed to §99 because we need for this generalization the apparatus of wave operators, which are dealt with in §§97 and 98.

To conclude the present section we establish some auxiliary results, which will be needed in §98.

First we obtain an equation connecting the functions $\varphi(t)$ and $\psi(t)$ in the representation (2).

It follows from (4) and (5) that

$$\int_{-\infty}^{\infty} \frac{\psi(\mu)\, d\sigma(\mu)}{\mu - \zeta} = \left[1 - \gamma \int_{-\infty}^{\infty} \frac{d\sigma(\mu)}{\mu - \zeta}\right] \int_{-\infty}^{\infty} \frac{\varphi(\lambda)\, d\rho(\lambda)}{\lambda - \zeta}. \tag{4'}$$

Passing to the limit in (4′) we obtain that — almost everywhere —

$$\pm \pi i \psi(\xi)\sigma'(\xi) + \int_{-\infty}^{\infty}{}' \frac{\psi(\mu)\, d\sigma(\mu)}{\mu - \xi} = \frac{1}{\omega(\xi \pm i0)} \int_{-\infty}^{\infty} \frac{\varphi(\lambda)\, d\rho(\lambda)}{\lambda - (\xi \pm i0)}, \tag{7}$$

where the prime after the first integral denotes that the Cauchy principal value is to be taken. By subtracting one of the relations in (7) from the other, we find that

$$2\pi i \psi(\xi)\sigma'(\xi) = \frac{1}{\omega(\xi + i0)} \int_{-\infty}^{\infty} \frac{\varphi(\lambda)\, d\rho(\lambda)}{\lambda - (\xi + i0)} - \frac{1}{\omega(\xi - i0)} \int_{-\infty}^{\infty} \frac{\varphi(\lambda)\, d\rho(\lambda)}{\lambda - (\xi - i0)}$$

or

$$2\pi i \psi(\xi)\sigma'(\xi) = \frac{\omega(\xi - i0) - \omega(\xi + i0)}{|\omega(\xi + i0)|^2} \int_{-\infty}^{\infty} \frac{\varphi(\lambda)\, d\rho(\lambda)}{\lambda - (\xi + i0)}$$

$$+ \frac{1}{\omega(\xi - i0)} \left\{ \int_{-\infty}^{\infty} \frac{\varphi(\lambda)\, d\rho(\lambda)}{\lambda - (\xi + i0)} - \int_{-\infty}^{\infty} \frac{\varphi(\lambda)\, d\rho(\lambda)}{\lambda - (\xi - i0)} \right\}.$$

Hence, by (6′), it follows that, almost everywhere,

$$2\pi i\psi(\xi)\sigma'(\xi)=\frac{2\pi i}{\omega(\xi-i0)}\varphi(\xi)\rho'(\xi)-2\pi i\gamma\sigma'(\xi)\int_{-\infty}^{\infty}\frac{\varphi(\lambda)\,d\rho(\lambda)}{\lambda-(\xi+i0)}.$$

After reduction we obtain the required relation

$$\psi(\xi)=\omega(\xi+i0)\varphi(\xi)-\gamma\int_{-\infty}^{\infty}\frac{\varphi(\lambda)\,d\rho(\lambda)}{\lambda-(\xi+i0)} \tag{8}$$

almost everywhere on the set on which $\sigma'(\xi)\neq 0$.

We shall also need the following

Lemma. *If* $f(\lambda)\in L^2(-\infty,\infty)$, *then*

$$\lim_{t\to\infty}\int_{-\infty}^{\infty}\left|\int_{-\infty}^{\infty}\frac{\exp[it(\mu-\lambda)]-1}{\mu-\lambda}f(\lambda)\,d\lambda-\int_{-\infty}^{\infty}\frac{f(\lambda)\,d\lambda}{\lambda-(\mu+i0)}\right|^2 d\mu=0.$$

Proof. We have

$$\frac{1}{\pi}\int_{-\infty}^{\infty}\frac{f(\lambda)\,d\lambda}{\lambda-(\mu+i0)}=if(\mu)+\tilde{f}(\mu),$$

where

$$\tilde{f}(\mu)=\frac{1}{\pi}\int_{-\infty}^{\infty}\frac{f(\lambda)}{\lambda-\mu}\,d\lambda.$$

Let $F(x)$ be the Fourier transform of the function $f(\lambda)$, so that

$$f(\mu)=\underset{A\to\infty}{\mathrm{l.i.m.}}\frac{1}{\sqrt{(2\pi)}}\int_{-A}^{A}F(x)\exp[i\mu x]\,dx.$$

Then, as is well-known (see, *e.g.*, Titchmarsh [124])

$$\tilde{f}(\mu)=\underset{A\to\infty}{\mathrm{l.i.m.}}\frac{i}{\sqrt{(2\pi)}}\int_{-A}^{A}F(x)\,\mathrm{sign}\,x\exp[i\mu x]\,dx$$

and so

$$\frac{1}{2\pi i}\int_{-\infty}^{\infty}\frac{f(\lambda)\,d\lambda}{\lambda-(\mu+i0)}=\underset{A\to\infty}{\mathrm{l.i.m.}}\frac{1}{\sqrt{(2\pi)}}\int_{-A}^{A}F(x)\exp[i\mu x]\,dx.$$

On the other hand, by the convolution theorem, when $t>0$,

$$\frac{1}{2\pi i}\int_{-\infty}^{\infty}\frac{\exp[it(\mu-\lambda)]-1}{\mu-\lambda}f(\lambda)\,d\lambda$$

$$=\frac{1}{2\pi i}\frac{1}{\sqrt{(2\pi)}}\int_{-\infty}^{\infty}F(x)\exp(i\mu x)\,dx\int_{-\infty}^{\infty}\frac{1-\exp(-ity)}{y}\exp(iyx)\,dy$$

$$=\frac{1}{\pi\sqrt{(2\pi)}}\int_{-\infty}^{\infty}F(x)\exp(i\mu x)\,dx\int_{0}^{\infty}\frac{\sin(yx)-\sin y(x-t)}{y}\,dy$$

$$=\frac{1}{\sqrt{(2\pi)}}\int_{0}^{t}F(x)\exp(i\mu x)\,dx.$$

Comparison of these relations proves the lemma.

§97. Definition and formal properties of wave operators

Let A and B be self-adjoint operators in H, let R(A) and R(B) be the sub-spaces formed by the regular elements of the operators A and B, and let P_A and P_B be the operators of orthogonal projection on to R(A) and R(B). We adopt the following

Definition. If the strong limit

$$U_+\{B, A\}=\lim_{t\to\infty}\exp(itB)\exp(-itA)P_A \tag{1}$$

exists, then the ordered pair of operators A, B is said to have a *wave operator*† $U_+\{B, A\}$.

Similarly the wave operator $U_-\{B, A\}$ is defined by

$$U_-\{B, A\}=\lim_{t\to-\infty}\exp(itB)\exp(-itA)P_A. \tag{2}$$

In this section we establish a number of properties of wave operators, assuming that these operators exist. Proofs of their existence will be given in §98. We shall discuss only the operator (1); the formulation and proofs of the properties of the operator (2) are exactly analogous.

Theorem 1. For any real τ,

$$\exp(i\tau B)U_+\{B, A\}=U_+\{B, A\}\exp(i\tau A). \tag{3}$$

† The concept of wave operators was first introduced by C. Møller by means of the expression $\lim_{\pm t\to\infty}\exp(itB)\exp(-itA)$. (See Møller [84]).

For,

$$\exp[i\tau B]U_+\{B, A\} = \lim_{t\to\infty} \exp[i(t+\tau)B]\exp[-i(t+\tau)A]\exp[i\tau A]P_A$$

$$= \lim_{t\to\infty} \exp[i(t+\tau)B]\exp[-i(t+\tau)A]P_A \exp[i\tau A] = U_+\{B, A\}\exp[i\tau A].$$

Theorem 2. For any real λ,

$$F_\lambda U_+\{B, A\} = U_+\{B, A\}E_\lambda, \tag{4}$$

where E_λ *and* F_λ *are the spectral functions of the operators* A *and* B *respectively.*

Proof. (3) implies that, for any $f, g \in \mathrm{H}$,

$$\int_{-\infty}^{\infty} \exp[i\tau\lambda]\, d(F_\lambda U_+\{B, A\}f, g) = \int_{-\infty}^{\infty} \exp[i\tau\lambda]\, d(U_+\{B, A\}E_\lambda f, g),$$

whence, by the uniqueness theorem for the Fourier–Stieltjes integral, we obtain the relation (4). Q.E.D.

Theorem 3. The operator $U_+\{B, A\}$ *is partially isometric. Its initial set is* $\mathrm{R}(A)$ *and its final set lies in* $\mathrm{R}(B)$.

Proof. For any $f \in \mathrm{H}$,

$$\|U_+\{B, A\}f\| = \lim_{t\to\infty} \|\exp[itB]\exp[-itA]P_A f\| = \|P_A f\|, \tag{5}$$

from which the proof of the first two assertions follows. To prove the third assertion, we observe that, by (4) and (5),

$$(F_\lambda U_+\{B, A\}f, U_+\{B, A\}) = \|F_\lambda U_+\{B, A\}f\|^2$$
$$= \|U_+\{B, A\}E_\lambda f\|^2 = \|P_A E_\lambda f\|^2 = \|E_\lambda P_A f\|^2 = (E_\lambda h, h) \tag{5'}$$

where $h = P_A f \in \mathrm{R}(A)$.
Since the right-hand side of (5′) is absolutely continuous, so is its left-hand side. Therefore the vector $U_+\{B, A\}f$ is a regular element of the operator B, *i.e.*, $U_+\{B, A\}f \in \mathrm{R}(B)$. Q.E.D.

Theorem 4. If the final set of the operator $U_+\{B, A\}$ *coincides with* $\mathrm{R}(B)$, *then the operators* A_a *and* B_a *are unitarily equivalent, and, for any* $f \in \mathrm{R}(A) \cap \mathrm{D}(A)$,

$$U_+^* B_a U_+ f = A_a f, \tag{6}$$

where $U_+ = U_+\{B, A\}$.

Proof. The first assertion will follow from Theorem 3 if the relation (6) is proved; so let us prove (6). To do this, we note that, by (5′), for any $f \in \mathrm{R}(A)$,

$$\int_{-\infty}^{\infty} \lambda^2 d(F_\lambda U_+ f, U_+ f) = \int_{-\infty}^{\infty} \lambda^2 d(E_\lambda f, f),$$

where both integrals either converge or both diverge together. So, if $f \in \mathrm{D}(A) \cap \mathrm{R}(A)$, then $U_+ f \in \mathrm{D}(B)$, and if $U_+ f \in \mathrm{D}(B)$ for $f \in \mathrm{R}(A)$, then $f \in \mathrm{D}(A)$. Further, by Theorem 2, for any $g \in \mathrm{H}$,

$$\int_{-\infty}^{\infty} \lambda d(F_\lambda U_+ f, g) = \int_{-\infty}^{\infty} \lambda d(U_+ E_\lambda f, g);$$

but this means that

$$B_{\mathrm{a}} U_+ f = U_+ A_{\mathrm{a}} f, \tag{6'}$$

which gives (6). (Q.E.D.

Theorem 5. (The multiplication theorem for wave operators)
If the wave operators $U_+\{B, A\}$ and $U_+(C, B)$ exist, then so does the wave operator $U_+\{C, A\}$, and

$$U_+\{C, A\} = U_+\{C, B\} U_+\{B, A\}. \tag{7}$$

Proof. By definition,

$$U_+\{B, A\} = \lim_{t \to \infty} \exp[itB] \exp[-itA] P_A,$$

$$U_+\{C, B\} = \lim_{t \to \infty} \exp[itC] \exp[-itB] P_B.$$

Therefore

$$U_+\{C, B\} U_+\{B, A\} = \lim_{t \to \infty} \exp[itC] \exp[-itB] P_B \exp[itB] \exp[-itA] P_A.$$

Replacing P_B by $I - (I - P_B)$, we find that

$$U_+\{C, B\} U_+\{B, A\} = \lim_{t \to \infty} \exp[itC] \exp[-itA] P_A$$
$$- \lim_{t \to \infty} \exp[itC] \exp[-itB](I - P_B) \exp[itB] \exp[-itA] P_A.$$

The first term on the right is $U_+\{C, A\}$. It remains to show that the second term is zero. But, as $t \to \infty$, the operator

$$(I - P_B) \exp[itB] \exp[-itA] P_A \to 0,$$

and, for any t, the operator

$$\exp[itC]\exp[-itB]$$

is unitary; hence their product tends to zero. Q.E.D.

Theorem 6. The final set of the operator $U_+\{B, A\}$ coincides with $\mathrm{R}(B)$ if and only if the operator $U_+\{A, B\}$ exists.

Proof. Sufficiency. If both the operators $U_+\{B, A\}$ and $U_+\{A, B\}$ exist, then, by Theorem 5, for any $f \in \mathrm{R}(B)$,

$$f = P_B f = U_+\{B, B\}f = U_+\{B, A\}U_+\{A, B\}f$$

or

$$f = U_+\{B, A\}g,$$

where $g = U_+\{A, B\}f$ belongs to $\mathrm{R}(A)$; this implies that f belongs to the final set of the operator $U_+\{B, A\}$.

Necessity. Suppose we know that the final set of the operator $U_+\{B, A\}$ is $\mathrm{R}(B)$; we have to show that, for any $f \in \mathrm{R}(B)$, the limit

$$\lim_{t \to \infty} \exp[itA]\exp[-itB]f$$

exists.

Given an element f, there is an element $g \in \mathrm{R}(A)$ such that

$$f = U_+\{B, A\}g.$$

Therefore

$$f = \lim_{t \to \infty} f_t, \qquad \text{where} \qquad f_t = \exp[itB]\exp[-itA]g.$$

But from the equality

$$\begin{aligned}\exp[itA]\exp[-itB]f &= \exp[itA]\exp[-itB](f - f_t + f_t)\\ &= \exp[itA]\exp[-itB](f - f_t) + g,\end{aligned}$$

it follows that

$$g = \lim_{t \to \infty} \exp[itA]\exp[-itB]f;$$

thus the operator $U_+\{A, B\}$ exists, and $g = U_+\{A, B\}f$. Q.E.D.

The operators $U_+\{A, B\}$ and $U_+\{B, A\}$ are clearly mutually inverse on the respective sets, and so

$$U_+\{B, A\} = U_+^*\{A, B\}. \tag{8}$$

We now introduce the additional concept of the *scattering operator* for the ordered pair A, B; this is the operator

$$S\{B, A\} = U_+^*\{B, A\}U_-\{B, A\}, \tag{9}$$

it being assumed that all the four operators $U_\pm\{B, A\}$, $U_\pm\{A, B\}$ exist.

Theorem 7. *The scattering operator $S\{B, A\}$ in the space $\mathrm{R}(A)$ is unitary and it commutes with the operator A.*

Proof. The first assertion follows from Theorem 6.

Further, we show that $S\{B, A\}$ commutes with the resolution of the identity E_λ for the operator A, for any $\lambda \in (-\infty, \infty)$. For brevity we write $U_\pm\{B, A\}$ as $U_\pm$. Then, from (9), using Theorem 2 twice, we have

$$S\{B, A\}E_\lambda = U_+^* U_- E_\lambda = U_+^* F_\lambda U_- = (F_\lambda U_+)^* U_-$$
$$= (U_+ E_\lambda)^* U_- = E_\lambda U_+^* U_- = E_\lambda S\{B, A\}.$$

This means, by the theorem in §80, that

$$S\{B, A\}A \subseteq AS\{B, A\},$$

i.e., A and $S\{B, A\}$ commute in H, and therefore they also commute in the sub-space $\mathrm{R}(A)$, which reduces both these operators. And if $f \in \mathrm{R}(A)$, then

$$(E_\lambda Sf, Sf) = (SE_\lambda f, Sf) = (E_\lambda f, f),$$

where $S = S\{B, A\}$; this means that f (in $\mathrm{R}(A)$) and Sf either both belong to $\mathrm{R}(A) \cap \mathrm{D}(A)$, or both do not belong to $\mathrm{R}(A) \cap \mathrm{D}(A)$. Therefore, on $\mathrm{R}(A)$,

$$A_a S\{B, A\} = S\{B, A\}A_a,$$

or, in the whole of H,

$$AS\{B, A\} = S\{B, A\}AP_A. \qquad \text{Q.E.D.}$$

The following theorem on the multiplication of scattering operators also holds.

Theorem 8. *If the scattering operators $S\{B, A\}$ and $S\{C, B\}$ exist, then the scattering operator $S\{C, A\}$ exists, and*

$$S\{C, A\} = S^+\{C, B\}S\{B, A\},$$

where

$$S^+\{C, B\} = U_+\{A, B\}S\{C, B\}U_+\{B, A\}.$$

Proof. The existence of the operator $S\{C, A\}$ follows from Theorem 5. Further, we have

$$\begin{aligned}S\{C, A\} &= U_+^*\{C, A\}U_-\{C, A\}\\ &= U_+^*\{B, A\}U_+^*\{C, B\}U_-\{C, B\}U_-\{B, A\}\\ &= U_+^*\{B, A\}U_+^*\{C, B\}U_-\{C, B\}U_+\{B, A\}U_+^*\{B, A\}U_-\{B, A\}\\ &= U_+\{A, B\}S\{C, B\}U_+\{B, A\}S\{B, A\}, \qquad \text{by (8).} \qquad \text{Q.E.D.}\end{aligned}$$

§98. Existence of wave operators in the case of finite-dimensional perturbations

This section links up directly with §96, the notation of which we shall continue to use here. We suppose initially, as in §96, that the operator K is 1-dimensional and that its range is generated by an element g, by means of which a separable Hilbert space H_g is constructed. We also assume that H_g coincides with the whole of H. In $\mathrm{H} = \mathrm{H}_g$ we define an operator V_+ by putting

$$V_+f = \int_{-\infty}^{\infty} \omega(\lambda + i0)\varphi(\lambda)\, dF_\lambda P_B g \tag{1}$$

if

$$f = \int_{-\infty}^{\infty} \varphi(\lambda)\, dE_\lambda g.$$

We also introduce the operator

$$U_t = \exp[itB]\exp[-itA]P_A \qquad (-\infty < t < \infty),$$

and we prove that

$$\lim_{t\to\infty} U_t f = V_+ f \qquad (f \in \mathrm{H}). \tag{2}$$

We observe that it suffices to prove this relation for $f \in \mathrm{R}(A)$, because, if $f \perp \mathrm{R}(A)$, then both $U_t f = 0$ and $V_+ f = 0$, the last result coming from the fact that, by (1) and the formula (6) in §96, we have, for any $f \in \mathrm{H}$,

$$\begin{aligned}\|V_+f\|^2 &= \int_{-\infty}^{\infty} |\omega(\lambda + i0)|^2 |\varphi(\lambda)|^2\, d(F_\lambda P_B g, g)\\ &= \int_{-\infty}^{\infty} |\varphi(\lambda)|^2 |\omega(\lambda + i0)|^2\, d\sigma_a(\lambda) = \int_{-\infty}^{\infty} |\varphi(\lambda)|^2\, d\rho_a(\lambda) = \|P_A f\|^2,\end{aligned} \tag{3}$$

where $\sigma_a(\lambda)=(F_\lambda P_B g, g)$ is the absolutely continuous part of the function $\sigma(\lambda)=(F_\lambda g, g)$, and $\rho_a(\lambda)=(E_\lambda P_A g, g)$ is the absolutely continuous part of the function $\rho(\lambda)=(E_\lambda g, g)$. Thus

$$V_+P_A = V_+, \qquad U_tP_A = U_t. \tag{3'}$$

So we pass on to the proof of the relation (2) for $f\in R(A)$. For this we use the equality

$$(U_tf, h)-(f, h)=i\gamma\int_0^t(\exp[-i\tau A]f, g)(\exp[i\tau B]g, h)\,d\tau, \quad (h\in H). \tag{4}$$

It is sufficient to verify this equality for $f\in D(A)\cap R(A)$; in that case it is obtained by integrating the identity

$$\begin{aligned}\frac{d}{d\tau}&(\exp[i\tau B]\exp[-i\tau A]f, h)\\ &=i(\exp[i\tau B]B\exp[-i\tau A]f, h)-i(\exp[i\tau B]A\exp[-i\tau A]f, h)\\ &=i(\exp[i\tau B]K\exp[-i\tau A]f, h)\\ &=i\gamma(\exp[i\tau B](\exp[-i\tau A]f, g)g, h)\\ &=i\gamma(\exp[-i\tau A]f, g)(\exp[i\tau B]g, h).\end{aligned}$$

If

$$f=\int_{-\infty}^{\infty}\varphi(\lambda)\,dE_\lambda g, \qquad h=\int_{-\infty}^{\infty}\kappa(\mu)\,dF_\mu g,$$

then the relation (4) may be re-written in the form

$$\begin{aligned}&(U_tf, h)-(f, h)\\ &=i\gamma\int_0^t\left\{\int_{-\infty}^{\infty}\exp[-i\tau\lambda]\varphi(\lambda)\,d\rho(\lambda)\right\}\left\{\int_{-\infty}^{\infty}\exp[i\tau\kappa]\overline{\kappa(\mu)}\,d\sigma(\mu)\right\}d\tau,\end{aligned}$$

where, as may easily be seen the order of integration can be changed, and so

$$(U_tf, h)-(f, h)=\gamma\iint_{-\infty}^{\infty}\frac{\exp[it(\mu-\lambda)]-1}{\mu-\lambda}\varphi(\lambda)\overline{\kappa(\mu)}\,d\rho(\lambda)\,d\sigma(\mu).$$

In addition to the condition $f\in R(A)$, we now assume that $h\in R(B)$. In that case, $d\rho(\lambda)$ and $d\nu(\mu)$ in the double integral on the right may be replaced by $\rho'(\lambda)\,d\lambda$ and $\sigma'(\mu)\,d\mu$. We further assume, to start with, that

$$\varphi(\lambda)\rho'(\lambda)\in L^2(-\infty,\infty), \qquad \kappa(\mu)\sigma(\mu)\in L^2(-\infty,\infty). \tag{5}$$

Then, by the lemma in §96,

$$\lim_{t\to\infty} (U_t f, h) = (f, h) + \gamma \int\!\!\int_{-\infty}^{\infty} \frac{\varphi(\lambda)\rho'(\lambda)}{\lambda - (\mu + i0)} \overline{\kappa(\mu)} \sigma'(\mu)\, d\lambda\, d\mu,$$

whence, by (8) of §96,

$$\lim_{t\to\infty} (U_t f, h)$$

$$= (f, h) + \int_{-\infty}^{\infty} \omega(\mu + i0)\varphi(\mu)\overline{\kappa(\mu)}\, d\sigma(\mu) - \int_{-\infty}^{\infty} \psi(\mu)\overline{\kappa(\mu)}\, d\sigma(\mu)$$

$$= \int_{-\infty}^{\infty} \omega(\mu + i0)\varphi(\mu)\overline{\kappa(\mu)}\, d\sigma(\mu),$$

and by (1)

$$\lim_{t\to\infty} (U_t f, h) = (V_+ f, h). \tag{6}$$

We have proved (6) under the assumption (5), but we need it for the case where

$$\varphi(\lambda)\sqrt{\rho'(\lambda)} \in \mathrm{L}^2(-\infty, \infty), \qquad \kappa(\mu)\sqrt{\sigma'(\mu)} \in \mathrm{L}^2(-\infty, \infty), \tag{5'}$$

i.e., where

$$\varphi(\lambda) \in \mathrm{L}^2_{\rho_a}(-\infty, \infty), \qquad \kappa(\mu) \in \mathrm{L}^2_{\sigma_a}(-\infty, \infty), \tag{5''}$$

because only in this case will the relation (6) have been proved for arbitrary $f \in \mathrm{R}(A)$ and $h \in \mathrm{R}(B)$. However, it suffices here to observe that, if $\Omega(\lambda)$ traverses the set of all step-functions with compact support, then the set of functions

$$\varphi(\lambda) = \begin{cases} \Omega(\lambda)/\sqrt{\rho'(\lambda)}, & \rho'(\lambda) \neq 0, \\ 0, & \rho'(\lambda) = 0, \end{cases}$$

is dense in $\mathrm{L}^2_{\rho_a}(-\infty, \infty)$ and satisfies the condition (5), and the set of functions

$$\kappa(\mu) = \begin{cases} \Omega(\mu)/\sqrt{\sigma'(\mu)}, & \sigma'(\mu) \neq 0, \\ 0, & \sigma'(\mu) = 0, \end{cases}$$

is dense in $\mathrm{L}^2_{\sigma_a}(-\infty, \infty)$ and also satisfies the condition (5).

Since the operator U_t is uniform in t, it follows from (6) and (3′) that

$$\lim_{t\to\infty} P_B U_t = V_+ \tag{7}$$

everywhere in H, in the sense of weak convergence.

We now show that this equality also holds in the sense of strong convergence, and hence that the factor P_B can be omitted from (7). For, the weak convergence implies that

$$\|V_+f\|^2 = \lim_{t\to\infty} (P_B U_t f, V_+ f)$$
$$\leqslant \underline{\lim}_{t\to\infty} \|V_+f\|\cdot\|P_B \exp[itB]\exp[-itA]P_A f\|$$

whence

$$\|V_+f\| = \underline{\lim}_{t\to\infty} \|P_B \exp[itB]\exp[-itA]P_A f\|$$
$$\leqslant \overline{\lim}_{t\to\infty} \|P_B \exp[itB]\exp[-itA]P_A f\| \leqslant \|P_A f\|.$$

But, by (3),

$$\|V_+f\| = \|P_A f\|,$$

and therefore

$$\|V_+f\| = \lim_{t\to\infty} \|P_B \exp[itB]\exp[-itA]P_A f\| = \|P_A f\|,$$

from which it follows that (7) holds in the sense of strong convergence. And since

$$\|P_A f\|^2 = \|P_B \exp[itB]\exp[-itA]P_A f\|^2$$
$$+\|(I-P_B)\exp[itB]\exp[-itA]P_A f\|^2,$$

it follows that

$$\lim_{t\to\infty} (I-P_B)\exp[itB]\exp[-itA]P_A f = 0$$

in the sense of strong convergence, and this, thanks to (7), means that

$$\lim_{t\to\infty} U_t f = V_+ f,$$

i.e., the relation (2) is proved.

Theorem. *If A is any self-adjoint operator in H, if K is a finite-dimensional self-adjoint operator, and if B = A + K, then the wave operators $U_\pm\{B, A\}$ and $U_\pm\{A, B\}$ exist and create an isometric relation between the absolutely continuous parts of the operators A and B.*

Proof. It suffices to give the proof for the operator $U_+\{B, A\}$.

If the operator K is 1-dimensional and if $H = H_g$, then the proof has already been given above. In the general case of a 1-dimensional operator

K when $\mathrm{H_g} \neq \mathrm{H}$, the wave operator

$$U_+\{B, A\} = \lim_{t\to\infty} \exp[itB]\exp[-itA]P_A$$

also exists, and, on the sub-space $\mathrm{H}\ominus\mathrm{R}(A)$, it is evidently the null operator; on the sub-space $\mathrm{R}(A)\cap(\mathrm{H}-\mathrm{H_g})$ it is the identity operator; and finally, on the sub-space $\mathrm{H_g}$ it may be defined, in accordance with (1) and (2), by the equality

$$U_+f = \int_{-\infty}^{\infty} \omega(\lambda + i0)\varphi(\lambda)\, dF_\lambda P_{(\mathrm{R_B}\cap\mathrm{H_g})}g$$

if

$$f = \int_{-\infty}^{\infty} \varphi(\lambda)\, dE_\lambda g.$$

The transition from a 1-dimensional perturbation to any finite-dimensional perturbation is effected by means of Theorem 5 in §97.

§99. Transition to the general case of nuclear perturbations

Lemma. (This delicate lemma is due to M. Rozenblum [111].)

Let A be a self-adjoint operator in H *and let* E_λ *be its spectral function. If the element* $f \in \mathrm{H}$ *is a regular element for A, and if almost everywhere*

$$\frac{d(E_\lambda f, f)}{d\lambda} \leq \mathcal{M}^2, \tag{1}$$

then, for any Hilbert–Schmidt operator T with absolute norm N(T), the inequality

$$\int_{-\infty}^{\infty} \|T\exp[-itA]f\|^2\, dt \leq 2\pi\mathcal{M}^2 N^2(T) \tag{2}$$

holds.

Proof. Let $\{g_n\}_1^\infty$ be an orthonormal basis in the range $\mathrm{R}(T)$. Then

$$\|T\exp[-itA]f\|^2 = \sum_{n=1}^{\infty} |(T\exp[-itA]f, g_n)|^2$$
$$= \sum_{n=1}^{\infty} \left|\int_{-\infty}^{\infty} \exp(-it\lambda)\, d(E_\lambda f, T^*g_n)\right|^2. \tag{3}$$

We shall obtain estimates for the integrals appearing in (3). To do this, we show that, by (1), and for any $h \in \mathrm{H}$, and in particular, for $h = T^*g_n$, the function $(E_\lambda f, h)$ is absolutely continuous, and that its derivative

$\dfrac{d(E_\lambda f, h)}{d\lambda}$ belongs to $L^2(-\infty, \infty)$, and satisfies the inequality

$$\int_{-\infty}^{\infty} \left|\frac{d(E_\lambda f, h)}{d\lambda}\right|^2 d\lambda \leq \mathcal{M}^2 \|h\|^2.$$

For, if H_f is the sub-space generated by the set of vectors $E(\Delta)f$ when Δ traverses the set of all intervals of the real line, then we can put

$$h = h_1 + h_2, \qquad \text{where} \quad h_1 \in H_f \quad \text{and} \quad h_2 \perp H_f.$$

Further, (see §83), there is a function $\varphi(\mu)$ such that

$$h_1 = \int_{-\infty}^{\infty} \varphi(\mu)\, dE_\mu f,$$

$$\|h_1\|^2 = \int_{-\infty}^{\infty} |\varphi(\mu)|^2\, d(E_\mu f, f) \leq \|h\|^2.$$

Therefore

$$(E_\lambda f, h) = (E_\lambda f, h_1) = \int_{-\infty}^{\lambda} \overline{\varphi(\mu)}\, d(E_\mu f, f),$$

from which our assertion follows.

This result shows that the integral

$$\int_{-\infty}^{\infty} \exp[-it\lambda]\, d(E_\lambda f, T^* g_n) = \int_{-\infty}^{\infty} \exp[-it\lambda] \frac{d(E_\lambda f, T^* g_n)}{d\lambda}\, d\lambda$$

can be regarded as a Fourier–Plancherel transform. The Parseval relation

$$\int_{-\infty}^{\infty} \left|\int_{-\infty}^{\infty} \exp[-it\lambda]\, d(E_\lambda f, T^* g_n)\right|^2 dt = 2\pi \int_{-\infty}^{\infty} \left|\frac{d(E_\lambda f, T^* g_n)}{d\lambda}\right|^2 d\lambda$$
$$\leq 2\pi \mathcal{M}^2 \|T^* g_n\|^2$$

therefore holds, and hence it follows that

$$\int_{-\infty}^{\infty} \sum_{n=1}^{m} \left|\int_{-\infty}^{\infty} \exp[-it\lambda]\, d(E_\lambda f, T^* g_n)\right|^2 dt \leq 2\pi \mathcal{M}^2 \sum_{n=1}^{m} \|T^* g_n\|^2$$
$$\leq 2\pi \mathcal{M}^2 N^2(T^*) = 2\pi \mathcal{M}^2 N^2(T).$$

By (3) and Fatou's theorem, we have, for any finite α, β with $\beta > \alpha$,

$$\int_{\alpha}^{\beta} \sum_{n=1}^{\infty} \left|\int_{-\infty}^{\infty} \exp[-it\lambda]\, d(E_\lambda f, T^* g_n)\right|^2 dt = \int_{\alpha}^{\beta} \|T \exp[-itA] f\|^2\, dt$$
$$\leq 2\pi \mathcal{M}^2 N^2(T),$$

and hence, letting $\alpha \to -\infty$ and $\beta \to +\infty$, we obtain (2). Q.E.D.

We now pass on to the generalization of the results of §§96 and 98 to the case of arbitrary nuclear perturbations K. The complete formulation of this generalization, including the propositions already established, is as follows.

Theorem. *Let A be any self-adjoint operator in H, and let K be a self-adjoint nuclear perturbation such that, in some orthonormal basis $\{g_i\}_1^\infty$,*

$$Kf = \sum_{j=1}^{\infty} \lambda_j (f, g_j) g_j \qquad (f \in \mathrm{H}), \tag{4}$$

where

$$\sum_{j=1}^{\infty} |\lambda_j| < \infty.$$

Then

1°. *the wave operator*

$$U_+ = U_+\{B, A\} = \lim_{t\to\infty} \exp[itB]\exp[-itA]P_{\mathrm{A}},$$

where $B = A + K$, exists.

2°. *It is partially isometric, with initial set $\mathrm{R}(A)$ and final set $\mathrm{R}(B)$.*

3°. *The absolutely continuous parts A_{a} and B_{a} of the operators A and B are unitarily equivalent:*

$$B_{\mathrm{a}} = U_+ A_{\mathrm{a}} U_+^*, \qquad A_+ = U_+^* B_{\mathrm{a}} U_+.$$

4°. *The other wave operators $U_-\{B, A\}$ and $U_\pm\{A, B\} = U_\pm^*\{B, A\}$ also exist and have similar properties.*

Proof. Thanks to the results of the previous sections, it suffices to prove just the first assertion. To do this, we put

$$K_m f = \sum_{j=1}^{m} \lambda_j (f, g_j) g_j, \qquad s_m = \sum_{j=1}^{m} |\lambda_j|, \qquad B_m = A + K_m,$$

$$U_t^{(m)} = \exp[itB_m]\exp[-itA]P_{\mathrm{A}} \qquad (-\infty < t < \infty)$$

for any finite m. Using, as in §98, the result of differentiating the vector

$$\exp[itB_m]\exp[-itA]f$$

with respect to t, we find that, for $f \in \mathrm{R}(A)$,

$$(U_t^{(m)} - U_s^{(m)})f = i \int_s^t \exp[i\tau B_m] K_m \exp[-i\tau A] f \, d\tau.$$

Since we have already proved the existence of the wave operator

$$U_+^{(m)} = U_+\{B_m, A\} \qquad (m<\infty),$$

we can let $t\to\infty$ in the last relation to obtain

$$(U_+^{(m)} - U_s^{(m)})f = i\int_s^\infty \exp[i\tau B_m]K_m \exp[-i\tau A]f\,d\tau. \tag{5}$$

Since

$$\|U_+^{(m)}f\| = \|f\| \quad \text{and} \quad \|U_s^{(m)}f\| = \|f\|,$$

we have

$$\|(U_+^{(m)} - U_s^{(m)})f\| = 2\mathcal{R}(((U_+^{(m)} - U_s^{(m)})f, U_+^{(m)}f)).$$

Hence, using (5) and Theorem 1 of §97 we find that

$$\|(U_+^{(m)} - U_s^{(m)})f\| = 2\mathcal{R}\left(i\int_s^\infty (K_m \exp[-i\tau A]f, U_+^{(m)} \exp[-i\tau A]f)\,d\tau\right). \tag{6}$$

We now express the perturbation K_m in the form

$$K_m = W_m\,|K_m| = W_m\,|K_m|^{1/2}\,|K_m|^{1/2}, \tag{7}$$

where $W_m = \operatorname{sign} K_m$ is a partially isometric operator. We assume initially that, for the element f,

$$\mathcal{M} = \operatorname{ess\,sup} \frac{d(E_\lambda f, f)}{d\lambda} < \infty. \tag{8}$$

Then (6) and (7) imply that

$$\|(U_+^{(m)} - U_s^{(m)})f\|^2 \leq 2\left[\int_s^\infty \||K_m|^{1/2} \exp[-i\tau A]f\|^2\,d\tau\right]^{1/2}$$
$$\times\left[\int_s^\infty \||K_m|^{1/2} W_m^* U_+^{(m)} \exp[-i\tau A]f\|^2\,d\tau\right]^{1/2}, \tag{9}$$

and both integrals exist by virtue of (8) and Rozenblum's lemma. We replace s by $-\infty$ in the second integral. By the same lemma and the estimate in §31 for the absolute norm of a product, which here gives

$$N^2(|K_m|^{1/2}\, W_m^* U_+^{(m)}) \leq s_m,$$

the inequality (9) can be expressed in the form

$$\|(U_+^{(m)} - U_s^{(m)})f\| \leq (8\pi\mathcal{M}^2 s_m)^{1/4}\left[\int_s^\infty \||K_m|^{1/2} \exp[-i\tau A]f\|^2\,d\tau\right]^{1/4}.$$

Hence it follows that

$$\|(U_t^{(m)}-U_s^{(m)})f\|\leqslant\|(U_+^{(m)}-U_t^{(m)})f\|+\|(U_+^{(m)}-U_s^{(m)})f\|$$

$$\leqslant(8\pi\mathcal{M}^2 s_m)^{1/4}\left\{\left[\int_s^\infty \||K_m|^{1/2}\exp[-i\tau A]f\|^2\,d\tau\right]^{1/4}\right. \tag{10}$$

$$\left.+\left[\int_t^\infty \||K_m|^{1/2}\exp[-i\tau A]f\|^2\,d\tau\right]^{1/4}\right\}.$$

We shall prove that in this inequality we can let $m\to\infty$. For, on the one hand, for any m,

$$s_m\leqslant s_\infty=\sum_{j=1}^\infty|\lambda_j|<\infty,$$

and, as can be seen from the representation (4),

$$\||K_m|^{1/2}h\|\leqslant\||K|^{1/2}h\| \qquad \text{for all } h\in \mathrm{H}.$$

On the other hand, as $m\to\infty$,

$$U_t^{(m)}h\to U_t h \qquad (h\in\mathrm{H}),$$

because, in the sense of strong convergence (at least),

$$\exp[itB_m]\to\exp[itB].$$

This follows from the fact that at each point of continuity of the spectral function F_λ of the operator B, by Theorem 2 of §79, the limit relation

$$\lim_{t\to\infty}F_\lambda^{(m)}=F_\lambda$$

holds (in the sense of strong convergence), $F_\lambda^{(m)}$ being the resolution of the identity for the operator B.

Thus, as $m\to\infty$, we find from (10) that

$$\|(U_t-U_s)f\|\leqslant(8\pi\mathcal{M}^2 s_\infty)^{1/4}\left\{\left[\int_s^\infty \||K|^{1/2}\exp[-i\tau A]f\|^2\,d\tau\right]^{1/4}\right.$$

$$\left.+\left[\int_t^\infty \||K|^{1/2}\exp[-i\tau A]f\|^2\,d\tau\right]^{1/4}\right\},$$

where

$$U_t=\exp[itB]\exp[-itA]P_{\mathrm{A}}.$$

It follows from this inequality that $\lim_{t\to\infty}U_t f$ exists for all elements $f\in\mathrm{R}(A)$ which satisfy the inequality (8). But the set of all such elements f for all possible $\mathcal{M}$ is dense in $\mathrm{R}(A)$, and the operator U_t is uniformly bounded with respect to t. Therefore

$$\lim_{t\to\infty} U_t f \text{ exists for all } f \in R(A),$$

and this means that the wave operator $U_+\{B, A\}$ exists. Q.E.D.

In conclusion we remark that the wave operators and scattering operators for a given pair of self-adjoint operators $\{A, B\}$ exist not only in the case of nuclear perturbations, *i.e.*, when the difference $B-A$ is nuclear, but also when the difference of some positive integral power of the resolvents of these operators, *i.e.*,

$$[R(z; A)]^p - [R(z; B)]^p \qquad (\mathcal{I}(z) \neq 0),$$

is nuclear, and also in a number of more general cases; see Birman and Krein [18], Birman [16, 17], and Kato [49, 50].

8

Theory of extensions of symmetric operators

§100. Deficiency indices

Let T be any linear operator; for the time being we do not even assume its domain to be dense in H.

A number λ (real or complex) is called a *point of regular type* for the operator T if there is a $k = k(\lambda) > 0$ such that, for all $f \in D(T)$,

$$\|(T - \lambda I)f\| \geq k\, \|f\|.$$

Therefore the eigenvalues of T are not points of regular type for T. Also, if λ is a point of regular type for T, then the operator $(T - \lambda I)^{-1}$ exists and is bounded, although its domain may not be dense in H; conversely, if $(T - \lambda I)^{-1}$ exists and is bounded, then λ is a point of regular type.

If λ_0 is a point of regular type, then, for $|\lambda - \lambda_0| \leq \delta \leq \frac{1}{2}k(\lambda_0)$ and for any $f \in D(T)$,

$$\begin{aligned}\|(T - \lambda I)f\| &\geq \|(T - \lambda_0 I)f\| - |\lambda - \lambda_0| \,.\, \|f\| \\ &\geq \{k(\lambda_0) - \delta\} \,.\, \|f\| \geq \tfrac{1}{2}k(\lambda_0) \,.\, \|f\|.\end{aligned}$$

This result, which has already been used in proving Theorem 4 in §48, shows that the set of points of regular type is always open. This set of points will be called the *field of regularity* of the operator T.

If A is a symmetric operator, and if $z = x + iy$ (with $y \neq 0$), then (*cf.* the proof of Theorem 3 in §48), for any $f \in D(A)$,

$$\|(A - zI)f\|^2 = \|(A - xI)f\|^2 + y^2\, \|f\|^2 \geq y^2\, \|f\|^2,$$

from which we see that each of the upper and lower halves of the z-plane is a connected component of the field of regularity of any symmetric operator.

The field of regularity of any isometric operator V also consists of two connected components, *viz.*, the interior and exterior of the unit circle; because if $|\zeta| < 1$,

$$\|(V - \zeta I)f\| \geq \|Vf\| - |\zeta| \,.\, \|f\| = (1 - |\zeta|) \,.\, \|f\|,$$

and similarly, if $|\zeta|>1$,

$$\|(V-\zeta I)f\|\geqslant|\zeta|\,.\,\|f\|-\|Vf\|=(|\zeta|-1)\,.\,\|f\|.$$

Theorem. *If Γ is a connected component of the field of regularity of a linear operator T, then the dimension of the sub-space* $\mathrm{H}\ominus\mathrm{R}(T-\lambda I)$ *is the same for all $\lambda\in\Gamma$.*

Proof. Let P_λ be the operator of orthogonal projection on to the sub-space

$$\mathrm{N}_\lambda=\mathrm{H}\ominus\mathrm{R}(T-\lambda I).$$

If we show that, for any $\lambda_0\in\Gamma$, there is a $\delta=\delta(\varepsilon)>0$ such that $|\lambda-\lambda_0|<\delta$ implies

$$\|P_\lambda-P_{\lambda_0}\|<1, \tag{1}$$

then it will follow from the theorem in §39 that

$$\dim \mathrm{N}_\lambda=\dim \mathrm{N}_{\lambda_0}.$$

Now, by the Heine–Borel lemma, any two points λ_1, λ_2 in Γ can be joined by a path which is covered by a finite number of these constructed δ-neighbourhoods; hence we obtain the result that

$$\dim \mathrm{N}_{\lambda_1}=\dim \mathrm{N}_{\lambda_2}.$$

So, let λ_0 be a fixed point of the domain Γ and let

$$\delta=\delta(\lambda_0)\leqslant\tfrac{1}{3}k(\lambda_0).$$

Since

$$k(\lambda_0)\,\|f\|\leqslant\|(T-\lambda_0 I)f\|\leqslant\|(T-\lambda I)f\|+|\lambda-\lambda_0|\,.\,\|f\|,$$

it follows that, if $|\lambda-\lambda_0|\leqslant\delta$, then

$$\|(T-\lambda I)f\|\geqslant\tfrac{2}{3}k(\lambda_0)\,\|f\|.$$

Let us assume that $|\lambda-\lambda_0|\leqslant\delta$; then, for any $h\in\mathrm{N}_\lambda$ with $\|h\|=1$,

$$\begin{aligned}\|(I-P_{\lambda_0})h\|&=\sup_{f\in\mathrm{D}(T)}\frac{|(h,(T-\lambda_0 I)f)|}{\|(T-\lambda_0)f\|}\\&=\sup_{f\in\mathrm{D}(T)}\frac{|(h,(T-\lambda I)f+(\lambda-\lambda_0)f|}{\|(T-\lambda_0 I)f\|}=\sup_{f\in\mathrm{D}(T)}\frac{|\lambda-\lambda_0|\,(h,f)}{\|(T-\lambda_0 I)f\|}\leqslant\frac{1}{2}\end{aligned} \tag{2}$$

and for any $h\in\mathrm{N}_{\lambda_0}$ $(\|h\|=1)$

$$\|(I-P_\lambda)h\|\leqslant 1/2, \tag{3}$$

By the definition of the aperture of two linear manifolds (§36), it follows from (2) and (3) that

$$\|P_\lambda - P_{\lambda_0}\| \leqslant 1/2 .$$

The theorem† has now been proved.

From these considerations it follows that

$$\lim_{\lambda \to \lambda_0} \|P_\lambda - P_{\lambda_0}\| = 0 \qquad (\lambda, \lambda_0 \in \Gamma).$$

This relation expresses the fact that the sub-space N_λ *rotates continuously* round the point $f=0$ as the point λ ranges over the domain Γ.

Let us agree that the *deficiency number* of a linear manifold M shall mean the dimension of its complement $\mathrm{N} = \mathrm{H} \ominus \mathrm{M}$:

$$\operatorname{def} \mathrm{M} = \dim \mathrm{N} .$$

The deficiency number may be finite or infinite.

The theorem just proved justifies the following

Definition. The deficiency number of the linear manifold

$$\mathrm{M}_\lambda = \mathrm{R}(T - \lambda I)$$

for points λ belonging to a given connected component of the field of regularity of an operator T is called the *deficiency number of the operator* T in this component of the field of regularity; and $\mathrm{N}_\lambda = \mathrm{H} \ominus \mathrm{M}_\lambda$ is called the *deficiency sub-space of the operator* T *for the point* λ, and any non-zero element of the deficiency sub-space is called a *deficiency element.*

Every symmetric operator A has two deficiency numbers, *viz.*, $\mathfrak{m}$ in the lower half-plane and $\mathfrak{n}$ in the upper half-plane:

$$\operatorname{def} \mathrm{R}(A - zI) = \begin{cases} \mathfrak{m}, & \mathscr{I}(z) < 0, \\ \mathfrak{n}, & \mathscr{I}(z) > 0. \end{cases}$$

Similarly any isometric operator V has two deficiency numbers:

$$\operatorname{def} \mathrm{R}(V - \zeta I) = \begin{cases} \mathfrak{m}, & (|\zeta| > 1), \\ \mathfrak{n}, & (|\zeta| < 1). \end{cases}$$

The deficiency numbers of a symmetric operator or of an isometric operator form an ordered pair of numbers $(\mathfrak{m}, \mathfrak{n})$ which is called the *deficiency index* of the operator.

† This theorem is due to M. G. Krein and M. A. Krasnoselsky (1947) (see Krein and Krasnoselsky [67]). For symmetric operators (see 1° and 2° below) it had been proved earlier by von Neumann, and the fact itself, expressed by von Neumann's theorem, had been discovered still earlier by H. Weyl for 2nd-order differential operators, and by Hellinger for Jacobi matrices (see von Neumann [92], and Weyl [126]).

If it is known in advance that the deficiency numbers are finite, then they will be denoted by italic letters m, n.

From the above theorem we have immediately these four propositions:

1°. *If* $\left\{\begin{matrix}\textit{a symmetric operator}\\ \textit{an isometric operator}\end{matrix}\right\}$ *has a point of regular type on the* $\left\{\begin{matrix}\textit{real axis}\\ \textit{unit circle}\end{matrix}\right\}$, *then its deficiency numbers are equal:* $\mathfrak{m}=\mathfrak{n}$.

2°. *If* A *is a symmetric operator, then any non-real number* z *is an eigenvalue for the adjoint operator* A^* *and is of multiplicity* $\mathfrak{m}$ *if* $\mathscr{I}(z)>0$, $\mathfrak{n}$ *if* $\mathscr{I}(z)<0$.

For, if $f\in D(A)$ and if $g\in N_{\bar{z}}$, then

$$(g, (A-\bar{z}I)f)=0,$$

whence

$$(A^*g-zg, f)=0, \qquad \text{or} \qquad A^*g=zg.$$

Thus z is an eigenvalue of the operator A^* and is of multiplicity $\dim N_{\bar{z}}$.

3°. *The deficiency numbers of an isometric operator* V *can be defined by the equalities*

$$\mathfrak{m}=\operatorname{def} D(V), \qquad \mathfrak{n}=\operatorname{def} R(V).$$

Only the first equality needs proof. It follows from the fact that, for any $\zeta\neq 0$,

$$R(V-\zeta I)=(V-\zeta I)D(V)=\left(\frac{1}{\zeta}V-I\right)D(V)$$

$$=\left(\frac{1}{\zeta}I-V^{-1}\right)VD(V)=\left(V^{-1}-\frac{1}{\zeta}I\right)D(V^{-1})=R\left(V^{-1}-\frac{1}{\zeta}I\right)$$

and therefore, for $|\zeta|>1$,

$$\mathfrak{m}=\operatorname{def} R(V-\zeta I)=\operatorname{def} R\left(V^{-1}-\frac{1}{\zeta}I\right)=\operatorname{def} R(V^{-1}-0I)$$

$$=\operatorname{def} R(V^{-1})=\operatorname{def} D(V).$$

4°. *If* A *is any symmetric operator in* H, *and if* B *is a bounded, self-adjoint operator in* H, *then the deficiency indices of* A *and* $A+B$ *are the same.*

To prove this it suffices to take, for a fixed non-real λ_0, a family of operators

$$A-\lambda_0 I+tB \qquad (0\leqslant t\leqslant 1),$$

and to show that the dimension of the sub-space $H \ominus R(A + tB - \lambda_0 I)$ does not depend on t, by using a method similar to the proof of the main theorem of this section.

§101. More about the Cayley transform

In §79 we introduced the Cayley transform V of a closed,† symmetric operator A by means of the formulae

$$\left.\begin{aligned}(A - \bar{z}I)h &= f,\\ (A - zI)h &= Vf,\end{aligned}\right\} \tag{1}$$

where z is any non-real number, and h traverses $D(A)$.

In this section we shall assume that $\mathscr{I}(z) > 0$.

The operator V is expressed in terms of the operator A by the formula

$$Vf = (A - zI)(A - \bar{z}I)^{-1}f;$$

and the domain $D(V)$ of the operator A is $R(A - \bar{z}I)$.

We also notice that, by (1),

$$\left.\begin{aligned}h &= \frac{(V - I)f}{\bar{z} - z},\\ Ah &= \frac{(\bar{z}V - zI)f}{\bar{z} - z},\end{aligned}\right\} \tag{2}$$

and therefore

$$Ah = (\bar{z}V - zI)(V - I)^{-1}h. \tag{2'}$$

A very important result needed later is that:

the deficiency index $(\mathfrak{m}, \mathfrak{n})$ *of the operator* A *is the same as the deficiency index of the operator* V.

For, by definition,

$$\mathfrak{m} = \operatorname{def} R(A - \bar{z}I).$$

But $\quad R(A - \bar{z}I) = D(V) \quad$ and so $\quad D(V) = \mathfrak{m}$.

Again, by definition,

$$\mathfrak{n} = \operatorname{def} R(A - zI)$$

† We know from §46 that every symmetric operator admits closure. So from now on, when discussing symmetric operators, we shall assume them to be closed.

and $\quad R(A - zI) = R(V)$;
therefore $\quad \operatorname{def} R(V) = \mathfrak{n}$.
It only remains to apply 3° of §100.

Theorem 1. *If V is an isometric operator, and if the manifold* $R(V-I)$ *is dense in* H, *then the operator A defined by* (2′) *is symmetric and V is its Cayley transform.*

Proof. Since $R(V-I)$ is dense in H, the inverse operator $(V-I)^{-1}$ exists, for otherwise 1 would be an eigenvalue of the operator V. But if

$$Vg = g \qquad (g \neq 0),$$

then, for any $f \in D(V)$,

$$(Vf - f, g) = (Vf, g) - (f, g) = (Vf, Vg) - (f, g) = 0,$$

i.e., $\quad g \perp R(V-I)$.

Since the operator $(V-I)^{-1}$ exists, the operator

$$A = (\bar{z}V - zI)(V-I)^{-1}$$

is well defined and its domain is dense in H. We show that it is a symmetric operator.

Let f, g be any elements of $D(A) = R(V-I)$:

$$\begin{aligned} f &= V\varphi - \varphi, \\ g &= V\psi - \psi, \end{aligned} \qquad \varphi, \psi \in D(V).$$

Then

$$Af = (\bar{z}V - zI)\varphi = \bar{z}V\varphi - z\varphi,$$

$$Ag = (\bar{z}V - zI)\psi = \bar{z}V\psi - z\psi,$$

and so

$$(Af, g) = (\bar{z}V\varphi - z\varphi, V\psi - \psi) = (z + \bar{z})(\varphi, \psi) - \bar{z}(V\varphi, \psi) - z(\varphi, V\psi),$$

$$(f, Ag) = (V\varphi - \varphi, \bar{z}V\psi - z\psi) = (z + \bar{z})(\varphi, \psi) - \bar{z}(V\varphi, \psi) - z(\varphi, V\psi).$$

Hence

$$(Af, g) = (f, Ag).$$

It is easily shown that

$$V = \frac{A - zI}{A - \bar{z}I},$$

and so V is the Cayley transform of A. Q.E.D.

From now on we shall, if two operators A, V are connected by the relations (1) and (2), say that A is the Cayley transform of V, and V is the Cayley transform of A.

As an immediate consequence of what has been proved we have

Theorem 2. If A_1, A_2 are symmetric operators and if V_1, V_2 are their Cayley transforms, then A_2 is an extension of A_1 if and only if V_2 is an extension of V_1.

Theorem 2 reduces the problem of symmetric extensions of a given operator A to the problem of isometric extensions of its Cayley transform V.

By §10, two closed linear manifolds F and G can be respectively the domain and the range of an isometric operator if and only if their dimensions are equal; so isometric extensions of the operator V can be obtained in the following way.

In the deficiency sub-spaces $\mathrm{H} \ominus \mathrm{D}(V)$ and $\mathrm{H} \ominus \mathrm{R}(V)$ we choose two sub-spaces F and G of equal dimension, and construct an arbitrary isometric operator V_1 having F as its domain and G as its range. We then define a linear operator $\tilde{V}$, with $\mathrm{D}(\tilde{V}) = \mathrm{D}(V) \oplus \mathrm{F}$ as its domain and $\mathrm{R}(\tilde{V}) = \mathrm{R}(V) \oplus \mathrm{G}$ as its range, by the formulae

$$\tilde{V}f = \begin{cases} Vf & \text{for } f \in \mathrm{D}(V), \\ V_1 f & \text{for } f \in \mathrm{F}. \end{cases}$$

Then $\tilde{V}$ is evidently an isometric extension of V_1, and by taking all possible choices of F, G, V_1, we obtain all the isometric extensions $\tilde{V}$ of the operator V and each extension once only.

To find some symmetric extension $\tilde{A}$ of an operator A, we have to take the Cayley transform V of A, find an isometric extension $\tilde{V}$ of V by the above method, and finally carry out a Cayley transform† on $\tilde{V}$ to get $\tilde{A}$. The formulae corresponding to this process will be given in §102.

From the above considerations it follows, in particular, that: an operator A is a maximal, symmetric (or self-adjoint) operator if and only if its Cayley transform V is a maximal, isometric (or unitary) operator.

Hence we have

Theorem 3. A symmetric operator is maximal if and only if one of its deficiency numbers is zero.

A symmetric operator is self-adjoint if and only if both its deficiency numbers are zero.

† The number 1 is not an eigenvalue of $\tilde{V}$, because the equality $\tilde{V}g = g$ implies that $(\tilde{V}g, Vf) = (g, Vf)$ for all $f \in \mathrm{D}(V)$, *i.e.*, $(g, f) = (g, Vf)$, or $(g, f - Vf) = 0$. The manifold $(I - V)\mathrm{D}(V) = \mathrm{D}(A)$ is dense in H; therefore $g = 0$.

The method of extension just described also gives

Theorem 4. *Any symmetric operator A with deficiency index* $(\mathfrak{m}, \mathfrak{n})$ *can always be extended into a maximal operator.*

If $\mathfrak{m} \neq \mathfrak{n}$, *then none of these extensions is self-adjoint;*
if $\mathfrak{m} = \mathfrak{n} < \infty$, *then any maximal extension of A is self-adjoint;*
but if $\mathfrak{m} = \mathfrak{n} = \infty$, *then some of the maximal extensions of A are self-adjoint, and some are not self-adjoint.*

§102. The von Neumann formulae

Theorem. *Let A be any symmetric operator, with domain* D(A), *and let* $N_{\bar{z}}$ *and* N_z $(\mathcal{I}(z) > 0)$ *be any pair of its deficiency sub-spaces. Then the domain* D(A*) *of the operator* A* *can be expressed as the direct sum of the three linear manifolds* D(A), $N_{\bar{z}}$, N_z:

$$D(A^*) = D(A) \oplus N_{\bar{z}} \oplus N_z .$$

Proof. Recalling that $N_{\bar{z}}$ is the eigenspace of the operator A^* corresponding to the eigenvalue z (proposition 2° of §100), we denote an element of $N_{\bar{z}}$ by g_z.

We shall prove that any element f of $D(A^*)$ can be expressed in the form

$$f = f_0 + g_z + g_{\bar{z}}, \tag{1}$$

where $f_0 \in D(A)$, $g_z \in N_{\bar{z}}$, $g_{\bar{z}} \in N_z$. It should be noted that, as well as (1), the formula (1′) will also hold:

$$A^*f = Af_0 + zg_z + \bar{z}g_{\bar{z}} . \tag{1′}$$

Let $f \in D(A^*)$. We resolve the element $A^*f - zf$ into its components in the orthogonal sub-spaces M_z and N_z:

$$A^*f - zf = (Af_0 - zf_0) + (\bar{z} - z)g_{\bar{z}} .$$

But $A^*g_{\bar{z}} = \bar{z}g_{\bar{z}}$; therefore

$$A^*(f - f_0 - g_{\bar{z}}) = z(f - f_0 - g_{\bar{z}}) .$$

Hence

$$f - f_0 - g_{\bar{z}} \in N_{\bar{z}}$$

i.e.,

$$f - f_0 - g_{\bar{z}} = g_z$$

or

$$f = f_0 + g_z + g_{\bar{z}}.$$

To complete the proof it remains to show that the representation (1) of each element $f \in D(A^*)$ is unique. Assuming the contrary, suppose that

$$f_0 + g_z + g_{\bar{z}} = 0. \tag{2}$$

Then, operating on (2) with A^*, we get

$$Af_0 + zg_z + \bar{z}g_{\bar{z}} = 0. \tag{2'}$$

Multiplying (2) by z and subtracting from (2′), we have

$$Af_0 - zf_0 + (\bar{z} - z)g_{\bar{z}} = 0,$$

whence, since the terms are orthogonal, $(\bar{z} - z)g_{\bar{z}} = 0$.
Similarly $\qquad (\bar{z} - z)g_z = 0$.
Hence $\qquad f_0 = g_{\bar{z}} = g_z = 0.$ Q.E.D.

We now find $\mathscr{I}((A^*f, f))$ for any $f \in D(A^*)$.
By (1), (1′), we have $f = f_0 + g$, where $g = g_z + g_{\bar{z}}$, and so

$$\begin{aligned}(A^*f, f) &= (A^*f_0 + A^*g, f_0 + g)\\ &= (Af_0, f_0) + (g, Af_0) + (Af_0, g) + (zg_z + \bar{z}g_{\bar{z}}, g_z + g_{\bar{z}}).\end{aligned}$$

The sum of the first three terms is real, and so

$$\mathscr{I}((A^*f, f)) = \mathscr{I}(z\,\|g_z\|^2 + \bar{z}\,\|g_{\bar{z}}\|^2 + [z(g_z, g_{\bar{z}}) + \bar{z}(g_{\bar{z}}, g_z)]),$$

and again the quantity in the square brackets is real, so that finally

$$\mathscr{I}((A^*f, f)) = \mathscr{I}(z\,\|g_z\|^2 - z\,\|g_{\bar{z}}\|^2). \tag{3}$$

In accordance with (3), the domain $D(A^*)$ consists of three (non-linear) manifolds:
Γ^+, the set of elements f for which $\mathscr{I}((A^*f, f)) > 0$,
Γ^-, the set of elements f for which $\mathscr{I}((A^*f, f)) < 0$,
and Γ°, the set of elements f for which (A^*f, f) is real.
An element

$$f = f_0 + g_z + g_{\bar{z}}$$

belongs to Γ^+, Γ^-, Γ° according as

$$\|g_z\| > \|g_{\bar{z}}\|, \qquad \|g_z\| < \|g_{\bar{z}}\|, \qquad \|g_z\| = \|g_{\bar{z}}\| \qquad (\text{if } \mathscr{I}(z) > 0).$$

We now find a representation similar to (1) for the domain $D(\tilde{A})$ of any symmetric extension $\tilde{A}$ of an operator A.

To emphasize the dependence of the sub-spaces F and G introduced in §101 on z, we write F_z and G_z. Thus $F_z \subseteq N_{\bar{z}}$, $G_z \subseteq N_z$.

From the considerations in §101 it follows that

$$D(\tilde{A}) = (\tilde{V} - I)D(\tilde{V}) = (\tilde{V} - I)(D(V) \oplus F_z)$$
$$= (V - I)D(V) \oplus (V_1 - I)F_z = D(A) \oplus (V_1 - I)F_z,$$

or, putting $V_1 = -V'$,

$$D(\tilde{A}) = D(A) \oplus (V' + I)F_z.$$

Since $A^* \supset \tilde{A}$, it follows that, if

$$f = f_0 + g_z + V'g_{\bar{z}} \qquad (g \in F_z), \tag{4}$$

then

$$\tilde{A}f = Af_0 + zg_z + \bar{z}V'g_z. \tag{4'}$$

We shall call (1) and (4) respectively the *first* and *second Neumann formulae* (see von Neumann [92]; in §§102–104 we have followed von Neumann's treatment).

From the first Neumann formula it follows immediately that the dimension of $D(A^*)$ modulo $D(A)$ is given by

$$\dim D(A^*)/D(A) = \dim N_{\bar{z}} + \dim N_z. \tag{5}$$

We point out that, by using the first Neumann formula and the manifolds Γ^+, Γ^-, Γ° introduced earlier, it is easy to derive the theorem about the constancy of the deficiency numbers for a symmetric operator in each of the half-planes $\mathscr{I}(z) \gtrless 0$ (*i.e.*, proposition 2 of §100), without using the general theorem in §100. In particular, when the deficiency numbers are equal, the result referred to follows from the one formula (5), since the left-hand side of (5) clearly does not depend on z.

The second Neumann formula together with the equality (4′) describes all symmetric extensions $\tilde{A}$ of a given operator A. If, in particular, the operator A has equal deficiency numbers, and if $\tilde{A}$ is a self-adjoint extension of A, then in the formula (4) the element g_z will traverse the whole sub-space $N_{\bar{z}}$ and $V'g_z$ the whole sub-space N_z.

Conversely, if in (4) the element g_z traverses the whole of $N_{\bar{z}}$ and $V'g_z$ the whole of N_z, then the operator $\tilde{A}$ will be a self-adjoint extension of the operator A.

If A has deficiency index (m, n) and its symmetric extension $\tilde{A}$ has deficiency index $(m-p, n-p)$ where $m, n < \infty$, then it follows from the second Neumann formula that

$$\dim D(\tilde{A})/D(A) = p.$$

We illustrate the general theory by taking as an example the operator of differentiation, P, which we introduced in §55. The equation

$$P^*g = zg$$

has the form

$$\frac{dg}{dt} + izg = 0,$$

and its formal solution is the function

$$g(t) = Ce^{-izt}. \tag{6}$$

In the case of the whole axis ($-\infty < t < \infty$), this function belongs to $L^2(-\infty, \infty)$ only if $C = 0$. Therefore the deficiency index of the operator P on the whole axis is (0, 0). In the case of the semi-axis ($0 \leqslant t < \infty$), the deficiency index is (0, 1), because the function (6) belongs to $L^2(0, \infty)$ if $\mathscr{I}(z) < 0$, but does not belong to $L^2(0, \infty)$ if $\mathscr{I}(z) > 0$. Finally, in the case of the interval ($0 \leqslant t \leqslant 2\pi$), the deficiency index will be (1, 1) since the function (6) belongs to $L^2(0, 2\pi)$ for any z.

By putting in the formula (4)

$$z = i, \quad g_z = e^t, \quad \text{and}$$

$$V' g_z = \theta g_{\bar{z}},$$

where $g_{\bar{z}} = e^{2\pi - t}$ and θ is a constant with $|\theta| = 1$, and letting $\arg \theta$ vary over the interval $[0, 2\pi)$, we obtain all self-adjoint extensions $\tilde{P}$ of P (on the interval $[0, 2\pi]$) in the following form:

$$\varphi(t) = \varphi_0(t) + \alpha(e^t + \theta e^{2\pi - t}),$$

$$\tilde{P}\varphi(t) = i\varphi_0'(t) + \alpha i(e^t - \theta e^{2\pi - t});$$

here $\varphi_0(0) = \varphi_0(2\pi) = 0$ and α is an arbitrary parameter.

It is easily verified that this agrees with the result of §55, the relation between the present θ and the parameter $\theta_0 = \theta$ of §55 being

$$\theta = \frac{\theta_0 - e^{2\pi}}{1 - e^{2\pi}\theta_0}.$$

§103. Simple symmetric operators

A symmetric operator (or an isometric operator) is said to be *simple* if there is no sub-space which reduces it and on which it would induce a self-adjoint operator (or a unitary operator).

A symmetric operator A is simple if and only if its Cayley transform V

is simple. The truth of this assertion follows from

Theorem 1. *A sub-space* G *reduces a symmetric operator if and only if it reduces the Cayley transform* V *of* A.

Proof. Let $h \in \mathrm{D}(A)$, so that

$$h = (V - I)f, \qquad Ah = (\bar{z}V - zI)f \qquad (f \in \mathrm{D}(V)).$$

Suppose the sub-space G reduces V, and let P denote the projection operator on to G; then

$$Pf \in \mathrm{D}(V), \qquad VPf = PVf.$$

Therefore

$$Ph = (V - I)Pf \in \mathrm{D}(A)$$

and

$$APh = A(V - I)Pf = (\bar{z}V - zI)Pf = P(\bar{z}V - zI)f = PAh.$$

This proves the *sufficiency* of the condition in the theorem.

Necessity. Suppose that $f \in \mathrm{D}(V)$, and therefore

$$f = (A - \bar{z}I)h, \qquad Vf = (A - zI)h \qquad (h \in \mathrm{D}(A)).$$

Assuming that the sub-space G reduces the operator A, we have

$$Pf = (A - \bar{z}I)Ph \in \mathrm{D}(V).$$

It remains to check that $VPf = PVf$. But

$$VPf = V(A - \bar{z}I)Ph = (A - zI)Ph = P(A - zI)h = PVf. \qquad \text{Q.E.D.}$$

If an isometric operator V is not simple —— and this means that it has unitary restrictions, then among these unitary restrictions there is a *maximal* one (in the sense that it is an extension of any other unitary restriction of V).

For, the maximal unitary restriction of the operator V is the restriction of V to the closed linear envelope G_0 of all those sub-spaces which reduce V and in which V induces unitary operators.

Similarly, the *maximal self-adjoint restriction* of a non-simple, symmetric operator can be defined.

Lemma. *Let* V *be an isometric operator with equal deficiency numbers, let* U_0 *be its maximal unitary restriction and* U *be any unitary extension of* V*. Then* $\mathrm{D}(U_0)$ *is the orthogonal complement of the linear envelope* L *of the sub-spaces* $U^k(\mathrm{H} \ominus \mathrm{D}(V)) \qquad (k \in \mathbf{Z})$.

Proof. We introduce the sub-space

$$G = H \ominus \bar{L}.$$

It is invariant under both U and U^{-1}, and therefore, by Theorem 3 of §47, it reduces U. Let U' be the restriction of U to G.

Since G is orthogonal to L, it is also, *a fortiori*, orthogonal to $H \ominus D(V)$ and therefore is included in $D(V)$:

$$G \subseteq D(V).$$

Similarly $\qquad G \subseteq R(V).$

Therefore G reduces V, and so U' is the unitary restriction of V to G, and hence

$$G \subseteq D(U_0). \tag{1}$$

But $\qquad D(U_0) \subseteq D(V),$

and this implies that $D(U_0)$ is orthogonal to the sub-space $H \ominus D(V)$. But since $U^k D(U_0) = U_0^k D(U_0) = D(U_0)$, it follows that $U^k D(U_0)$ is orthogonal to $H \ominus D(V)$ for $k \in \mathbf{Z}$. Consequently $D(U_0)$ is orthogonal to L, *i.e.*,

$$D(U_0) \subseteq G. \tag{2}$$

Comparison of (1) and (2) proves the Lemma.

Later we shall need this immediate consequence of the lemma.

Corollary. In a non-separable space there are no simple symmetric operators which have equal and finite deficiency numbers.

Another consequence of the lemma is

Theorem 2. In order that an isometric operator V be simple it is necessary that for every unitary extension U of V, and it is sufficient that for some one of its unitary extensions U, the linear envelope of the sub-spaces $U^k(H \ominus D(V)) \quad (k \in \mathbf{Z})$ *shall be dense in* H.

From this theorem and Theorem 1 of §83 it follows that an arbitrary unitary (or self-adjoint) extension of a simple isometric (or symmetric) operator with deficiency index (1, 1) has a simple spectrum.

The following general result is proved similarly.

Theorem 3. The multiplicity of the spectrum of any unitary (or self-adjoint) extension of a simple isometric (or symmetric) operator with finite deficiency index (m, m) does not exceed m.

§104. The structure of maximal operators

Let H be a separable space, and $\{e_k\}_1^\infty$ some orthonormal basis in it. We consider linear operators V_+ and V_- defined by

$$V_+e_k = e_{k+1} \qquad (k=1,2,3,\dots),$$
$$V_-e_k = e_{k-1} \qquad (k=2,3,4,\dots).$$

It is clear that V_+ and V_- are isometric operators with the deficiency indices (0, 1) and (1, 0) respectively.

Since we want to use the Cayley transforms A_+ and A_- of the operators V_+ and V_-, we verify that the manifolds $R(V_+-I)$ and $R(V_--I)$ are dense in H. We establish this fact for the first manifold, for example. Taking

$$f = e_k + \frac{r-1}{r}\, e_{k+1} + \cdots + \frac{1}{r}\, e_{k+r-1} \qquad (k \in \mathbf{N}),$$

we find that

$$\begin{aligned}(V_+-I)f &= \left(e_{k+1} + \frac{r-1}{r}\, e_{k+2} + \cdots + \frac{1}{r}\, e_{k+r}\right) \\ &\quad - \left(e_k + \frac{r-1}{r}\, e_{k+1} + \cdots + \frac{1}{r}\, e_{k+r-1}\right) \\ &= \frac{1}{r}(e_{k+1} + e_{k+2} + \cdots + e_{k+r}) - e_k.\end{aligned}$$

But

$$\left\| \frac{1}{r}(e_{k+1} + e_{k+2} + \cdots + e_{k+r}) \right\|^2 = \frac{1}{r},$$

from which we conclude that the basis vectors e_k $(k \in \mathbf{N})$ are the limits of vectors in $R(V_+-I)$, and therefore the manifold $R(V_+-I)$ is dense in H.

The symmetric operators A_+ and A_- are maximal and have the deficiency indices (0, 1) and (1, 0) respectively.

Theorem 1. *The operators A_+ and A_- are irreducible.*

To prove this assertion for A_+ it suffices to show (*cf.* §103) that the operator V_+ is irreducible.

We assume, on the contrary, that a sub-space F (and therefore its orthogonal complement G) reduce V_+, and we put $V_+F = F_+$, $V_+G = G_+$.

Then

$$F_+ \subseteq F \quad \text{and} \quad G_+ \subseteq G.$$

The equality sign cannot hold simultaneously in both these relations, since $F_+ = F$ and $G_+ = G$ would imply that $V_+H = H$. Suppose then that, $F \ominus F_+ \neq \{0\}$, and let

$$f \in F \ominus F_+ \qquad (f \neq 0).$$

As an element of F, the vector f is orthogonal to G, and therefore

$$f \perp F_+ \oplus G_+. \tag{1}$$

(1) implies that $f \perp e_k \quad (k = 2, 3, \ldots)$, *i.e.*, $f = \alpha e_1$ with $\alpha \neq 0$; hence $e_1 \in F \ominus F_+$, and *a fortiori*, $e_1 \in F$.

Since F reduces V_+, the sub-space F contains all the e_k $(k = 2, 3, \ldots)$ as well as e_1, *i.e.*, $F = H$. But by the definition in §45, this means that V_+ is irreducible. The second part of the theorem is proved similarly.

Q.E.D.

We have thus shown that the operators A_+ and A_- are irreducible, and so they are simple symmetric operators.

The fundamental importance of these operators flows from the following

Theorem 2. If a simple symmetric operator A in a space H *has the deficiency index* (0, 1) *(resp.* (1, 0)*), then the space* H *is separable, and the operator A is isomorphic to the operator* A_+ *(resp.* A_-*).*

Proof. For definiteness, suppose that the deficiency index of the operator A is (0, 1). Let V be the Cayley transform of the operator A, and let $e_1 \perp R(V)$ with $\|e_1\| = 1$. We form the closed linear envelope M of the vectors $V^k e_1 = e_{k+1} \quad (k = 0, 1, 2, \ldots)$. M is a sub-space, and $\{e_k\}_0^\infty$ is an orthonormal basis in it.

Since M is invariant under V and V^{-1}, M reduces V, and therefore $H \ominus M$ also reduces V. Clearly, the restriction of V to M is isomorphic to V_+, and hence it has the deficiency index (0, 1). The restriction of V to $H \ominus M$ must be unitary, since otherwise at least one of the deficiency numbers of V would exceed the corresponding deficiency number of the operator V_+, and therefore that of the operator A.

But a simple operator V has no unitary restrictions. So $H \ominus M = \{0\}$, and the operator V is isomorphic to V_+, and consequently its Cayley transform A is isomorphic to A_+, as we had to show.

By means of the operator A_+ (or A_-) the maximal operator with deficiency index $(0, \mathfrak{n})$ (or $(\mathfrak{m}, 0)$) can be constructed. To do this, it suffices to construct the orthogonal sum of separable Hilbert spaces H_α, where α traverses a set of cardinality $\mathfrak{n}$(or $\mathfrak{m}$), and to set up the operators A_+ (or A_-) in each of them.

As J. von Neumann first showed, it is possible in this way to obtain any simple maximal operator; in other words, the following theorem holds.

Theorem 3. *A simple symmetric operator A which has the deficiency index* $(0, \mathfrak{n})$ *(or* $(\mathfrak{m}, 0)$*) can be split into an orthogonal sum of operators* A_+ *(or* A_-*):*

i.e., $A = \sum_{\alpha \in M} \oplus A_+^{(\alpha)}$ *(where M is a set of cardinality* $\mathfrak{n}$*)*

(or $A = \sum_{\alpha \in M} \oplus A_-^{(\alpha)}$ *(where M is a set of cardinality* $\mathfrak{m}$*)).*

Proof. Let V be the Cayley transform of the operator A, and let

$$D(V) = H, \qquad H \ominus R(V) = M_1 \qquad (\dim M_1 = \mathfrak{n}).$$

We put

$$V^k M_1 = M_{k+1} \qquad (k = 0, 1, 2, \ldots).$$

Then, as in Theorem 2, we obtain that

$$M_i \perp M_k \qquad (i \neq k;\ i, k \in \mathbf{N}),$$

and that the closure of the linear envelope of the sub-spaces M_k $(k \in \mathbf{N})$ coincides with H.

Let $\{e_1^{(\alpha)}\}_{\alpha \in M}$ be an orthonormal basis in M_1, and let $H^{(\alpha)}$ be the closure of the linear envelope of the vectors $\{e_k^{(\alpha)}\}_{k=1}^{\infty}$, where $e_{k+1}^{(\alpha)} = V^k e_1^{(\alpha)}$. The sub-spaces $H^{(\alpha)}$ and $H^{(\alpha')}$ are orthogonal if $\alpha \neq \alpha'$, and the orthogonal sum of all such sub-spaces is equal to H, since for each k the set $\{e_k^{(\alpha)}\}_{\alpha \in M}$ forms an orthonormal basis in M_k.

Clearly, each of the sub-spaces $H^{(\alpha)}$ reduces V, and the restriction of V to $H^{(\alpha)}$ is isomorphic to V_+. Passing over to the Cayley transforms completes the proof of the theorem. Q.E.D.

Theorem 3 provides the complete answer to the question concerning the structure of simple maximal operators.

We shall call A_+ and A_- (or V_+ and V_-) *elementary maximal operators.*

By means of the elementary maximal operators we can construct an operator with an assigned deficiency index $(\mathfrak{m}, \mathfrak{n})$. To do this, by the method described above, we have to construct operators with deficiency indices $(\mathfrak{m}, 0)$ and $(0, \mathfrak{n})$, and then form their orthogonal sum. However, it is not possible in general to construct an arbitrary *simple* operator with deficiency index $(\mathfrak{m}, \mathfrak{n})$ from elementary maximal operators.

As an illustration of the propositions of this section, we shall again consider the operator of differentiation, P, on the half-line $(0, \infty)$. This operator has the deficiency index $(0, 1)$. To elucidate the structure of the operator P, we go over to its Cayley transform.

$$V = (P - iI)(P + iI)^{-1},$$

and consider the powers $V^k g$, where $g = e^{-t}$ is the deficiency element of the operator P for the point i. We shall regard this element e^{-t} as a Chebyshev–Laguerre function of zero order (see §12) with doubled argument:

$$e^{-t} = \psi_0(2t);$$

and we prove that

$$V\psi_k(2t) = \psi_{k+1}(2t) \qquad (k = 0, 1, 2, \dots).$$

We express $\psi_k(2t)$ in the form

$$\psi_k(2t) = if'(t) + if(t) \equiv (P + iI)f;$$

since $f(0) = 0$, we have

$$f(t) = -ie^{-t} \int_0^t \psi_k(2s)e^s\, ds.$$

Therefore

$$V\psi_k(2t) = (P - iI)f = \psi_k(2t) - 2e^{-t} \int_0^t \psi_k(2s)e^s\, ds.$$

Thus it all comes down to proving the identity

$$\psi_{k+1}(2t) - \psi_k(2t) - 2e^{t} \int_0^t \psi_k(2s)e^s\, ds.$$

The proof presents no difficulty.

Since the Chebyshev–Laguerre functions form a complete orthonormal system in $L^2(0, \infty)$, the operator V is isomorphic to V_+, and so P is isomorphic to A_+. Thus we have proved that:

any simple operator with deficiency index $(0, 1)$ *or* $(1, 0)$ *is isomorphic to the operator of differentiation P on the half-line* $(0, \infty)$ *or* $(-\infty, 0)$ *respectively.*

The operator of differentiation P on the whole line $(-\infty, \infty)$ is isomorphic to some self-adjoint extension of the operator $A_+ \oplus A_-$. From this fact (*cf.* §89) the unitary equivalence of the operator P and the operator of multiplication by the independent variable can be deduced; this equivalence has already been established directly in §55.

§105. The spectra of self-adjoint extensions of a given symmetric operator

According to the general definition in §48, the spectrum of an operator which is symmetric but not self-adjoint contains the complement of the set of points of regular type for the operator in question. Although the spectrum is not indeed exhausted by this complement (thus, for example, at least one of the open half-planes $\mathscr{I}(z)>0$ or $\mathscr{I}(z)<0$ belongs to the residual spectrum), it does nevertheless occupy a special place in the spectrum of the operator.

Definition. The complement of the set of points of regular type of a symmetric operator is called the *spectral kernel* of this operator.

Since we want to classify the points which constitute the spectral kernel of a symmetric operator A, let us agree, as before, to denote by $A'_\lambda = A'$ the restriction of the operator to the sub-space $\mathrm{H} \ominus \mathrm{G}_\lambda$, where G_λ is either the eigen-manifold of A corresponding to λ if λ is an eigenvalue, or G_λ is the null space if λ is not an eigenvalue.

In the spectral kernel we distinguish firstly the *point part*, that is, the set of all eigenvalues of the operator; this set is empty if the operator is simple.

Turning now to the nature of the remaining part of the spectral kernel of the operator A, we observe that the operator $(A'-\lambda I)$ has an inverse for any λ. The set of all those values of λ for which the operator $(A'-\lambda I)^{-1}$ is unbounded clearly belongs to the spectral kernel, and we shall call this set the *continuous part* of the spectral kernel.† Thus every point of the spectral kernel belongs either to the point part, or the continuous part, or both.

For a self-adjoint operator, the concepts of a 'regular point' and a 'point of regular type' are the same. So the spectral kernel of a self-adjoint operator is the same as the spectrum of this operator. Consequently, the spectral kernel of a self-adjoint operator cannot be the empty set. But for an arbitrary symmetric operator, the last statement would be false (see the example in the footnote on p. 372).

If $\tilde{A}$ is a symmetric (or, in particular, a maximal or self-adjoint) extension of an operator A, then, as may easily be seen, the spectral kernel of $\tilde{A}$ contains the spectral kernel of A.

We note that there is one particular case in which the continuous part of the spectral kernel of A remains unchanged for symmetric extensions of

† It is easily seen that the continuous part of the spectral kernel of a symmetric operator A belongs to its continuous spectrum $\mathscr{C}(A)$ as defined in §93.

the operator. This happens when the deficiency numbers of A are finite. For, in this case — by the second Neumann formula (see §102) — if $\tilde{A}$ is any symmetric extension of the operator A, the manifold $(\tilde{A}'-\lambda I)\mathrm{D}(\tilde{A})$ is bigger than the manifold $(A'-\lambda I)\mathrm{D}(A)$ only by a finite number of dimensions (in the sense of dimension of the first manifold modulo the second), and so the operator $(\tilde{A}'-\lambda I)^{-1}$ is bounded when the operator $(A'-\lambda I)^{-1}$ is bounded.

From what has been said the following simple theorem follows.

Theorem 1. *All self-adjoint extensions of an operator with equal and finite deficiency numbers have the same continuous spectrum.*

The following theorem about the point part of the spectral kernel holds.

Theorem 2. *For an arbitrary extension of an operator with a deficiency index (m, m) into a self-adjoint operator the multiplicity of the eigenvalues is increased by not more than m; in particular, the multiplicity of any new eigenvalue does not exceed m.*

(Recall that use of an italic m implies that $m<\infty$; see p. 350)

Proof. Let $\tilde{A}$ be a self-adjoint extension of A, and let λ be an eigenvalue of multiplicity p, of the operator A, which becomes an eigenvalue of multiplicity $p+q$ of the operator $\tilde{A}$, and suppose that, contrary to this theorem, $q>m$. We choose a linearly independent set of solutions $f_1, f_2, \ldots, f_p, f_{p+1}, \ldots, f_{p+q}$ of the equation $\tilde{A}f-\lambda f=0$ such that $f_k \in \mathrm{D}(A)$ for $k \leq p$. Since the dimension of $\mathrm{D}(\tilde{A})$ modulo $\mathrm{D}(A)$ is equal to m, there are constants α_k such that

$$\alpha_1 f_{p+1}+\alpha_2 f_{p+2}+\cdots+\alpha_q f_{p+q} \in \mathrm{D}(A).$$

But this implies that the multiplicity of λ as an eigenvalue of the operator A exceeds p, contrary to hypothesis. Hence the theorem.

The following is, in a sense, a converse of Theorem 2.

Theorem 3. *If λ is a real point of regular type for a symmetric operator A with deficiency index (m, m), then there is a self-adjoint extension $\tilde{A}$ of A such that λ is an eigenvalue of multiplicity m for $\tilde{A}$.*

Proof. Let N_λ denote the linear manifold of all solutions of the equation

$$A^*g-\lambda g=0.$$

By the theorem on the invariance of a deficiency number in the field of regularity (*cf.* §100), the dimension of the manifold N_λ is equal to m.

The domain $D(A)$ of the operator A and the linear manifold N_λ are linearly independent, for otherwise λ would be an eigenvalue of A.

We put

$$D = D(A) \oplus N_\lambda, \tag{1}$$

and let $\tilde{A}$ denote the operator which coincides with the operator A on $D = D(\tilde{A})$, so that λ will be an eigenvalue of multiplicity m for $\tilde{A}$.

We shall prove that the operator $\tilde{A}$ is self-adjoint.

To do this, it suffices to prove that $\tilde{A}$ is symmetric, because it follows from (1) that

$$\dim D(\tilde{A}) = m \qquad (\operatorname{mod} D(A)).$$

If $f, g \in D(\tilde{A})$ and if

$$f = f_1 + f_2 \qquad (f_1 \in D(A),\ f_2 \in N_\lambda)$$

$$g = g_1 + g_2 \qquad (g_1 \in D(A),\ g_2 \in N_\lambda),$$

then

$$(\tilde{A}f, g) = (Af_1, g_1) + (A^*f_2, g_1) + (Af_1, g_2) + (A^*f_2, g_2)$$
$$= (Af_1, g_1) + \lambda(f_2, g_1) + \lambda(f_1, g_2) + \lambda(f_2, g_2)$$

and

$$(f, \tilde{A}g) = (f_1, Ag_1) + (f_1, A^*g_2) + (f_2, Ag_1) + (f_2, A^*g_2)$$
$$= (f_1, Ag_1) + \lambda(f_1, g_2) + \lambda(f_2, g_1) + \lambda(f_2, g_2),$$

from which it follows that the operator $\tilde{A}$ is symmetric. Q.E.D.

In conclusion we mention another theorem, on the number of solutions of the equation

$$A^*g - \lambda g = 0$$

when λ is real.

Theorem 4. *If A is a symmetric operator with deficiency index (m, m), and if λ is a real number which does not belong to the point spectrum of A, then the number $m(\lambda)$ of solutions of the equation*

$$A^*g - \lambda g = 0 \tag{2}$$

does not exceed the deficiency number m.

For the proof it suffices to construct by means of the manifold N_λ of solutions of equation (2) the domain $D(\tilde{A})$ from the formula (1), where again $\tilde{A} \subseteq A^*$.

From the proof of Theorem 3 it follows that the operator $\tilde{A}$ is a symmetric extension of A, and hence

$$m(\lambda) = \dim D(\tilde{A}) \leqslant m \qquad (\text{mod}\, D(A)). \qquad \text{Q.E.D.}$$

§106. M. G. Krein's formula for the resolvents of self-adjoint extensions of a given symmetric operator

In this section we consider symmetric operators with equal and finite deficiency numbers.

Let A_1 and A_2 be two self-adjoint extensions of an operator A which has the deficiency index (m, m). Thus

$$A_1 \supset A, \qquad A_2 \supset A.$$

It is natural to call every operator C which satisfies the conditions

$$A_1 \supset C, \qquad A_2 \supset C \tag{1}$$

a *common part* of the operators A_1 and A_2.

Clearly, among the operators C which satisfy the conditions (1) there is one which is an extension of *any* common part of the operators A_1 and A_2, and we shall call this operator the *maximal common part* of the operators A_1 and A_2. The maximal common part is either an extension of the operator A or it coincides with A; in the latter case, we shall say that the extensions A_1 and A_2 are *relatively prime.*

Two extensions A_1 and A_2 are relatively prime if and only if the conditions

$$h \in D(A_1), \qquad h \in D(A_2), \qquad A_1 h = A_2 h \tag{2}$$

together imply that $\quad h \in D(A)$.

If $p \quad (0 \leqslant p \leqslant m)$ is the maximum number of vectors which are linearly independent modulo $D(A)$ and which satisfy (2), then the deficiency index of the maximal common part A_0 of the operators A_1 and A_2 is $(m-p, m-p)$. In this case the operators A_1 and A_2 can be regarded as relatively prime, self-adjoint extensions of the operator A_0.

Our problem in this section is to derive a formula linking the resolvents of two self-adjoint extensions of an operator A.

Let $\mathring{B}$ be a fixed self-adjoint extension of A, and B be an arbitrary self-adjoint extension of A; let $\mathring{R}(z) = R(z; \mathring{B})$ and $R(z) = R(z; B)$ be their resolvents. Also let λ be any common regular point of the operators $\mathring{B}$ and B (in particular, λ can be an arbitrary non-real number).

In order not to exclude the case where $\mathring{B}$ and B are not relatively prime extensions of the operator A, we shall regard them as relatively prime extensions of their maximal common part A_0, which has the deficiency index (r, r) with $0<r\leqslant m$.

We put $\mathrm{M}_\lambda = \mathrm{R}(A_0-\lambda I)$ and $\mathrm{N}_\lambda = \mathrm{H}\ominus\mathrm{M}_\lambda$; then we have for the difference of the resolvents

$$(\mathring{R}(\lambda)-R(\lambda))f \begin{cases} =0 & \text{for } f\in\mathrm{M}_\lambda, \\ \in\mathrm{N}_{\bar\lambda} & \text{for } f\in\mathrm{N}_\lambda. \end{cases} \tag{3}$$

The latter result follows from the fact, for any $h\in\mathrm{M}_{\bar\lambda}$

$$\begin{aligned}(\{\mathring{R}(\lambda)-R(\lambda)\}f, h) &= (f, \{\mathring{R}(\lambda)-R(\lambda)\}^*h) \\ &= (f, \{\mathring{R}(\bar\lambda)-R(\bar\lambda)\}h) = (f, 0) = 0.\end{aligned}$$

We select any r linearly independent vectors $g_1(\bar\lambda), \ldots, g_r(\bar\lambda)$ from N_λ, and r linearly independent vectors $g_1(\lambda), \ldots, g_r(\lambda)$ from $\mathrm{N}_{\bar\lambda}$. Then, for any $f\in\mathrm{H}$, it follows from (3) that

$$(\mathring{R}(\lambda)-R(\lambda))f = \sum_{k=1}^{r} c_k g_k(\lambda). \tag{4}$$

By (4) the constants c_k are linear functionals of f, and we can put

$$c_k = (f, h_k(\lambda)) \qquad (k=1, 2, \ldots, r).$$

Since, by (3) and the linear independence of the vectors $g_1(\lambda), \ldots, g_r(\lambda)$,

$$(f, h_k(\lambda)) = 0 \qquad (k=1, 2, \ldots, r) \quad \text{for any } f\perp\mathrm{N}_\lambda,$$

it follows that

$$h_k(\lambda)\in\mathrm{N}_\lambda \qquad (k=1, 2, \ldots, r),$$

i.e.,

$$h_k(\lambda) = \sum_{i=1}^{r} \overline{p_{ik}(\lambda)} g_i(\bar\lambda) \qquad (k=1, 2, \ldots, r) \tag{5}$$

and (4) takes the form

$$(\mathring{R}(\lambda)-R(\lambda))f = \sum_{i,k=1}^{r} p_{ik}(\lambda)(f, g_i(\bar\lambda))g_k(\lambda). \tag{6}$$

We observe that the matrix function $[p_{ik}(\lambda)] = \mathscr{P}(\lambda)$, which is defined on the set of regular points of the operators $\mathring{B}$ and B, is non-singular. For, if $\det[p_{ik}(\lambda_0)]$ were zero, then, by (5), the vectors $h_k(\lambda_0)$ $(k=1,2,\ldots,r)$ would be linearly dependent, and this would imply that there is a non-zero vector h satisfying the conditions

$$h \perp h_k(\lambda_0), \qquad h \in \mathrm{N}_{\lambda_0} \qquad (k=1,2,\ldots,r).$$

We should then have, from (4),

$$(\mathring{R}(\lambda_0) - R(\lambda_0))h = 0,$$

and this contradicts the relative primeness of the operators $\mathring{B}$ and B as extensions of the operator A_0.

By omitting the element f from (6), and regarding the expressions $(\cdot, g_i(\bar{\lambda}))g_k(\lambda)$ $(i, k=1,2,\ldots,r)$ as operators, we obtain, for any value of λ in the set of regular points common to the operators $\mathring{B}$ and B, the formula

$$R(\lambda) = \mathring{R}(\lambda) - \sum_{i,k=1}^{r} p_{ik}(\lambda)(\cdot, g_i(\bar{\lambda}))g_k(\lambda). \tag{7}$$

Up to now the choice of the vector-functions $g_k(\lambda)$ and $g_i(\lambda)$ $(i, k=1,2,\ldots,r)$ has been left arbitrary. Further, in formula (6) the left-hand side, and so also the right-hand side, are regular analytic vector-functions of λ. We shall show that *the* $g_k(\lambda)$ $(k=1,2,\ldots,r)$ *can be determined as regular analytic vector-functions of* λ, *and we shall obtain a formula for the matrix function* $\mathcal{P}(\lambda)$ *corresponding to this choice.*

To do this, we choose some fixed valuc λ_0 and we introduce an opcrator

$$U(\lambda;\lambda_0) = (\mathring{B} - \lambda_0 I)(\mathring{B} - \lambda I)^{-1} = I + (\lambda - \lambda_0)\mathring{R}(\lambda)$$

with the domain

$$(\mathring{B} - \lambda I)\mathrm{D}(\mathring{B}) = \mathrm{H}$$

and the range

$$(\mathring{B} - \lambda_0 I)\mathrm{D}(\mathring{B}) = \mathrm{H}.$$

The operator $U(\lambda;\lambda_0)$ is defined by the formulae

$$(\mathring{B} - \lambda I)f = h, \qquad (f \in \mathrm{D}(\mathring{B})),$$
$$(\mathring{B} - \lambda_0 I)f = U(\lambda;\lambda_0)h,$$

from which it follows that the mapping from H to H determined by $U(\lambda;\lambda_0)$ is one-to-one.

In the particular case when $\lambda = \bar{\lambda}_0$, the operator $U(\lambda; \lambda_0)$ reduces to the Cayley transform of the operator $\mathring{B}$, and it maps the deficiency sub-space $N_{\bar{\lambda}_0}$ of the operator A_0 into its deficiency sub-space N_{λ_0}. We shall prove that in the general case

$$U(\lambda; \lambda_0)N_{\bar{\lambda}_0} = N_{\bar{\lambda}}.$$

We choose an arbitrary basis $g_1(\lambda_0), g_2(\lambda_0), \ldots, g_r(\lambda_0)$ — in general these vectors $g_k(\lambda_0)$ are not orthogonal nor normalized — and we show that

$$U(\lambda; \lambda_0)g_k(\lambda_0) \in N_{\bar{\lambda}}.$$

We have

$$\begin{aligned} A_0^*U(\lambda; \lambda_0)g_k(\lambda_0) &= A_0^*\{I + (\lambda - \lambda_0)\mathring{R}(\lambda)\}g_k(\lambda_0) \\ &= \lambda_0 g_k(\lambda_0) + (\lambda - \lambda_0)\mathring{B}\mathring{R}(\lambda)g_k(\lambda_0) \\ &= \lambda_0 g_k(\lambda_0) + (\lambda - \lambda_0)(I + \lambda\mathring{R}(\lambda))g_k(\lambda_0) \\ &= \lambda\{I + (\lambda - \lambda_0)\mathring{R}(\lambda)\}g_k(\lambda_0) = \lambda U(\lambda; \lambda_0)g_k(\lambda_0), \end{aligned}$$

i.e., $$U(\lambda; \lambda_0)g_k(\lambda_0) \in N_{\bar{\lambda}} \qquad (k = 1, 2, \ldots, r).$$

Moreover, since the mapping produced by $U(\lambda; \lambda_0)$ is one-to-one, the vectors $U(\lambda; \lambda_0)g_k(\lambda_0)$ form a basis in $N_{\bar{\lambda}}$, and we can take the vectors $g_k(\lambda)$ to be defined at any regular point of the operator $\mathring{B}$ by the formulae

$$g_k(\lambda) = U(\lambda; \lambda_0)g_k(\lambda_0) = g_k(\lambda_0) + (\lambda - \lambda_0)\mathring{R}(\lambda)g_k(\lambda_0), \qquad (k = 1, 2, \ldots, r),$$

and therefore they are regular analytic vector-functions of λ.

By means of the functional equation for the resolvent it is easy to verify that in such a case the relations

$$g_k(\mu) = U(\mu; \lambda)g_k(\lambda) = g_k(\lambda) + (\mu - \lambda)\mathring{R}(\mu)g_k(\lambda) \tag{8}$$

hold for any two points λ, μ of regularity of the operator $\mathring{B}$.

The value of the matrix-function $\mathscr{P}(\lambda)$ at any point λ which is regular for both $\mathring{B}$ and B is now determined from its value $\mathscr{P}(\lambda_0)$; to find the relevant formula we use the functional equation for the resolvent

$$R(\lambda) = R(\lambda_0) - (\lambda - \lambda_0)R(\lambda)R(\lambda_0). \tag{9}$$

But, by (7),

$$\left.\begin{aligned} R(\lambda) &= \mathring{R}(\lambda) - \sum_{i,k=1}^{r} p_{ik}(\lambda)(\cdot, g_i(\bar{\lambda}))g_k(\lambda), \\ R(\lambda_0) &= \mathring{R}(\lambda_0) - \sum_{i,k=1}^{r} p_{ik}(\lambda_0)(\cdot, g_i(\bar{\lambda}_0))g_k(\lambda_0) \end{aligned}\right\}. \tag{10}$$

Substituting from (10) into (9) (and using the equation like (9) for $\mathring{R}(\lambda)$), we find

$$
\begin{aligned}
&-\sum_{i,k=1}^{r} p_{ik}(\lambda)(\cdot, g_i(\bar{\lambda}))g_k(\lambda)\\
&=-\sum_{i,k=1}^{r} p_{ik}(\lambda_0)(\cdot, g_i(\bar{\lambda}_0))g_k(\lambda_0)\\
&\quad-(\lambda-\lambda_0)\sum_{i,k=1}^{r} p_{ik}(\lambda_0)(\cdot, g_i(\bar{\lambda}_0))\mathring{R}(\lambda)g_k(\lambda_0) \qquad (11)\\
&\quad-(\lambda-\lambda_0)\sum_{i,k=1}^{r} p_{ik}(\lambda)(\mathring{R}(\lambda_0)\cdot, g_i(\bar{\lambda}))g_k(\lambda)\\
&\quad+(\lambda-\lambda_0)\sum_{i,k,j,s=1}^{r} p_{ik}(\lambda)(g_s(\lambda_0), g_i(\bar{\lambda}))p_{js}(\lambda_0)(\cdot, g_j(\bar{\lambda}_0))g_k(\lambda).
\end{aligned}
$$

By using (8) we can express the sum of the second and third terms on the right as

$$
\begin{aligned}
&\sum_{i,k=1}^{r} p_{ik}(\lambda_0)(\cdot, g_i(\bar{\lambda}_0))\{g_k(\lambda_0)-g_k(\lambda)\}\\
&\quad+\sum_{i,k=1}^{r} p_{ik}(\lambda)(\cdot, g_i(\bar{\lambda}_0)-g_i(\bar{\lambda}))g_k(\lambda),
\end{aligned}
$$

and then (11) gives

$$
\begin{aligned}
&-\sum_{i,k=1}^{r} p_{ik}(\lambda_0)(\cdot, g_i(\bar{\lambda}_0))g_k(\lambda)+\sum_{i,k=1}^{r} p_{ik}(\lambda)(\cdot, g_i(\bar{\lambda}_0))g_k(\lambda)\\
&\quad+(\lambda-\lambda_0)\sum_{i,k,j,s=1}^{r} p_{ik}(\lambda)(g_s(\lambda_0), g_i(\bar{\lambda}))p_{js}(\lambda_0)(\cdot, g_j(\bar{\lambda}_0))g_k(\lambda)=0.
\end{aligned}
$$

Hence, since the vectors $g_k(\lambda)$ are linearly independent,

$$
\begin{aligned}
&-\sum_{i=1}^{r} p_{ik}(\lambda_0)(\cdot, g_i(\bar{\lambda}_0))+\sum_{i=1}^{r} p_{ik}(\lambda)(\cdot, g_i(\bar{\lambda}_0))\\
&\quad+(\lambda-\lambda_0)\sum_{i,j,s=1}^{r} p_{ik}(\lambda)(g_s(\lambda_0), g_i(\bar{\lambda}))p_{js}(\lambda)(\cdot, g_j(\bar{\lambda}_0))=0,
\end{aligned}
$$

and further, since the vectors $g_i(\bar{\lambda}_0)$ are linearly independent,

$$
-p_{ik}(\lambda_0)+p_{ik}(\lambda)+(\lambda-\lambda_0)\sum_{j,s=1}^{r} p_{is}(\lambda_0)(g_s(\lambda_0), g_j(\bar{\lambda}))p_{jk}(\lambda)=0,
$$

or, in matrix form

$$
\mathcal{P}(\lambda)-\mathcal{P}(\lambda_0)+(\lambda-\lambda_0)\mathcal{P}(\lambda_0)[(g_s(\lambda_0), g_j(\bar{\lambda}))]_{s,j=1}^{r}\mathcal{P}(\lambda)=0.
$$

Multiplying this on the right by $\mathcal{P}^{-1}(\lambda)$ and on the left by $\mathcal{P}^{-1}(\lambda_0)$, we finally obtain the required relation:

$$\mathcal{P}^{-1}(\lambda)=\mathcal{P}^{-1}(\lambda_0)+(\lambda-\lambda_0)[(g_s(\lambda_0),\ g_j(\bar{\lambda}))]_{j,s=1}^{r}. \tag{12}$$

It is easily verified that (12) implies that, for any two regular points λ and μ common to the two operators $\mathring{B}$ and B,

$$\mathcal{P}^{-1}(\lambda)=\mathcal{P}^{-1}(\mu)+(\lambda-\mu)[(g_s(\mu),\ g_j(\bar{\lambda}))]_{j,s=1}^{r}.$$

§107. Self-adjoint extensions of semi-bounded operators

Without loss of generality we can regard a semi-bounded operator as being a positive operator. Then the negative semi-axis belongs to its field of regularity, since, if $(Af, f)\geqslant 0$ and $\lambda<0$, then

$$\|(A-\lambda I)f\|^2=\|Af\|^2-2\lambda(Af, f)+\lambda^2\|f\|^2\geqslant\lambda^2\|f\|^2,$$

i.e.,
$$\|(A-\lambda I)f\|\geqslant|\lambda|\cdot\|f\|.$$

From this fact it follows, by proposition 1° in §100, that the deficiency numbers of a semi-bounded operator are equal, and so a semi-bounded operator has self-adjoint extensions.

Note that from the fact that the negative semi-axis belongs to the domain of regularity of a symmetric operator A it cannot be concluded that this operator is positive,† unless it is self-adjoint as well. If A is self-adjoint, then, for any $f\in D(A)$,

$$(Af, f)=\int_{-\infty}^{\infty} t\, d(E_t f, f)=\int_0^{\infty} t\, d(E_t f, f)\geqslant 0.$$

The square of any self-adjoint operator is a positive operator. Conversely, any positive self-adjoint operator A can be expressed as the square of some self-adjoint operator B. For, if

$$A=\int_0^{\infty} t\, dE_t,$$

then we can put, *e.g.*,

$$B=\int_0^{\infty}\sqrt{t}\, dE_t.$$

If a semi-bounded operator has finite deficiency numbers, then any

† For example, the field of regularity for the operator of differentiation on a finite interval is the whole plane, but this operator is not semi-bounded.

self-adjoint extension of this operator is also semi-bounded. Moreover, the following theorem holds.

Theorem 1. *If A is a positive operator with deficiency index (m, m) then any self-adjoint extension of A has only a finite number of negative eigenvalues, and the sum of their multiplicities does not exceed m.*

Proof. Let $\tilde{A}$ be a self-adjoint extension of the operator A, and let

$$\tilde{A} = \int_{-\infty}^{\infty} t\, dE_t.$$

Let Δ denote the interval $[-N, -\varepsilon)$.

To prove the theorem it suffices to show that, for any N, ε with $N > \varepsilon > 0$, the dimension of the sub-space $E(\Delta)\mathrm{H}$ does not exceed the deficiency number m.

Suppose, on the contrary, that, for some $N > \varepsilon > 0$,

$$\dim E(\Delta)\mathrm{H} > m. \tag{1}$$

Since $E(\Delta) \subseteq \mathrm{D}(\tilde{A})$ and $\dim \mathrm{D}(\tilde{A}) = m \pmod{\mathrm{D}(A)}$, it follows from (1) that there is in the sub-space $E(\Delta)\mathrm{H}$ a vector $f_0 \neq 0$ which also belongs to $\mathrm{D}(A)$. Then, for this f_0,

$$(Af_0, f_0) = (\tilde{A}f_0, f_0) = \int_{-N}^{-\varepsilon} t\, d(E_t f_0, f_0) < 0,$$

and this contradicts the fact that A is a positive operator. Q.E.D.

From the above considerations it is easy to prove also the following proposition, which is similar to Theorem 1 in §105.

Theorem 2. *If the negative part of the spectrum of one self-adjoint extension of an operator with deficiency index (m, m) is exhausted by a finite number of eigenvalues of finite multiplicities, then the same is true for any other self-adjoint extension of the given operator.*

Note that this Theorem 2 and also Theorem 1 of §105 are also immediate consequences of Theorems 1 and 2 respectively in §82.

The following theorem also holds for self-adjoint extensions of semi-bounded operators having arbitrary deficiency indices.

Theorem 3. *An operator A which is semi-bounded below can be extended into a self-adjoint operator $\tilde{A}$ having the same lower bound.* ($\tilde{A}$ is often called the *Friedrichs extension* of A.)

This theorem was stated as a conjecture by J. von Neumann and was proved later in various ways by different authors.†

We shall return to the proof of Theorem 3 later, but now we prove a simpler theorem, which was established by von Neumann.

Theorem 3′. *A semi-bounded operator A with lower bound μ can be extended into a self-adjoint operator $\tilde{A}$ with a lower bound which is not less than an arbitrarily chosen number $\mu' < \mu$.*

Proof. We first prove the theorem for $\mu = 1$ and $\mu' = 0$.

We have

$$(f, f) \leqslant (Af, f) \leqslant \|Af\| \cdot \|f\|,$$

i.e.,

$$\|Af\| \geqslant \|f\|,$$

from which it follows that the operator A^{-1} exists and is bounded on the sub-space $R(A)$. Repeating the argument of proposition 2° in §100, we easily see that

$$N_0 = H \ominus R(A)$$

is the eigen-manifold of the operator A^* corresponding to the eigenvalue $\lambda = 0$. Moreover, the sub-space N_0 and the manifold $D(A)$ are linearly independent, for if an element g of N_0 belonged to $D(A)$, then would $Ag = 0$, and this is possible only if $g = 0$.

We define an extension $\tilde{A}$ of the operator A on to the domain

$$D(\tilde{A}) = D(A) \oplus N_0$$

by putting

$$\tilde{A}h = Af$$

for $h = f + g, \qquad f \in D(A), \qquad g \in N_0$.

$\tilde{A}$ is clearly a symmetric extension of A, because, if

$$h_1, h_2 \in D(\tilde{A}) \qquad (h_i = f_i + g_i, \quad f_i \in D(A), \quad g_i \in N_0, \quad i = 1, 2),$$

then

$$(\tilde{A}h_1, h_2) = (Af_1, f_2 + g_2) = (Af_1, f_2)$$
$$= (f_1, Af_2) = (f_1 + g_1, Af_2) = (h_1, \tilde{A}h_2).$$

† See Stone [119], Friedrichs [29], and Freudental [28]. Not long ago Kilpi produced a new proof of Theorem 3, which, in contrast to the constructions of the authors just cited, remains within the framework of the von Neumann theory of extensions; see Kilpi [52].

A very complete theory of self-adjoint extensions of semi-bounded operators, preserving the norm, was constructed by M. G. Krein [60].

Further, it is clear that the sub-space N_0 reduces $\tilde{A}$, because $N_0 \subset D(\tilde{A})$ and $\tilde{A}h = 0$ for $h \in N_0$. The orthogonal complement $R(A)$ of N_0 also reduces $\tilde{A}$. Let $\tilde{A}'$ and $\tilde{A}''$ be the restrictions of $\tilde{A}$ to $R(A)$ and N_0 respectively.

The range $R(\tilde{A}')$ of the operator $\tilde{A}'$ fills the sub-space $R(A)$, and therefore (see §46) the operator $\tilde{A}'$ is self-adjoint. And since $\tilde{A}'' = 0$, the operator

$$\tilde{A} = \tilde{A}' \oplus \tilde{A}''$$

also is self-adjoint.

Further, for

$$h = f + g \qquad (f \in D(A),\ g \in N_0),$$

we have

$$(\tilde{A}h, h) = (Af, f+g) = (Af, f) \geq \|f\|^2 > 0,$$

i.e., the lower bound of the operator $\tilde{A}$ is not less than 0.

Thus the theorem has been proved for the case $\mu = 1$, $\mu' = 0$, and the general case is reduced to this one by the linear transformation

$$A_1 = \frac{1}{\mu - \mu'} A - \frac{\mu'}{\mu - \mu'} I. \qquad \text{Q.E.D.}$$

We now return to Theorem 3. It is easily proved in the case where the deficiency numbers are finite.

For, let A be a symmetric operator with a lower bound $\mu > -\infty$ and with finite deficiency numbers, so that $\dim N_{\bar{z}} = \dim N_z < \infty$. By Theorem 3′, for every $\mu_n < \mu$, there is a self-adjoint extension $\tilde{A}_n$ of the operator A with a lower bound $\geq \mu_n$. By Neumann's formulae (4) and (4′) in §102, to each such extension will correspond an isometric mapping V'_n of the whole of $N_{\bar{z}}$ on to the whole of N_z. As $\mu_n \to \mu$, because the sequence $\{V'_n\}_1^\infty$ of finite-dimensional isometric operators is compact, there is a sub-sequence $\{V'_{n_k}\}$ such that $V'_{n_k} \to V'$, where V' is some isometric mapping of the whole of $N_{\bar{z}}$ on to the whole of N_z. From formulae (4), (4′) of § 102 again, it is clear that the operator V' so obtained defines the required self-adjoint extensions $\tilde{A}$ of the operator A, preserving its lower bound.

The general case of Theorem 3 will be dealt with in §109, which gives an account of the main results of M. G. Krein in the theory of extensions of symmetric operators semi-bounded below with preservation of their lower bound. Here we give a brief general description of M. G. Krein's method.

First of all we observe that, to prove Theorem 3, it suffices to show that any positive symmetric operator admits a positive self-adjoint extension. This follows from the fact that the lower bound of the operator $A-\mu I$ is zero if the lower bound of A is μ, and conversely. If then, the operator A is positive, then the point $\lambda=-1$ is one of its points of regular type, and so the bilinear transformation

$$S=(I-A)(I+A)^{-1} \tag{2}$$

is meaningful.

The operator S defined on the sub-space $D(S)=(I+A)D(A)$ satisfies the condition for symmetry

$$(Sg_1, g_2)=(g_1, Sg_2) \qquad (g_1, g_2 \in D(S))$$

and, as can easily be proved, its norm does not exceed 1. Further, it turns out that the problem of finding a self-adjoint extension of the operator A, preserving positivity, is equivalent to constructing a self-adjoint extension $\tilde{S}$ of S, satisfying the condition $\|\tilde{S}\| \leqslant 1$.

M. G. Krein's construction for this purpose and some of his auxiliary results needed to describe all the possible extensions referred to in Theorem 3 are given in the next section.

§108. Norm-preserving, self-adjoint extensions of a bounded symmetric operator whose domain is not dense in H

Let S be an arbitrary linear operator whose domain $D(S)$ is not dense in H and which satisfies the symmetry condition

$$(Sf, g)=(f, Sg) \qquad \text{for any } f, g \in D(S).$$

We shall call such operators† *symmetric operators with non-dense domains.*

Theorem 1. *Any bounded symmetric operator S whose domain $D(S) \subset H$ is not dense in* H *can be extended into a self-adjoint operator $\tilde{S}$ $(D(\tilde{S})=H)$, preserving the norm, i.e. with $\|\tilde{S}\|=\|S\|$.*

Proof. Without loss of generality we may take $\|S\|=1$. We shall also take the domain $D(S)$ to be closed, for, if it is not, then the operator S can be extended by continuity on to the closure of $D(S)$. Let P be the

† The general theory of extensions of such operators is similar to the von Neumann theory of extensions, which we have expounded; it was developed by M. A. Krasnoselsky (see Krasnoselsky [57]).

ortho-projector on to D(S); we construct the operators

$$S_1 = PS, \qquad S_2 = (I-P)S,$$

so that $\quad D(S_1) = D(S_2) = D(S), \quad$ and, for any $\varphi \in D(S)$,

$$S\varphi = S_1\varphi + S_2\varphi. \tag{1}$$

We shall extend each of the operators S_1 and S_2 separately on to H.

To extend S_1 we write down, for any $f, g \in D(S)$, the equality

$$(Sf, g) = (f, Sg) = (Pf, Sg) = (f, S_1 g). \tag{2}$$

The left-hand side has a meaning with any $g \in H$, and for each such g it represents a linear functional of $f \in D(S_1)$, with norm $\leqslant \|g\|$. Since D(S) is a Hilbert space, by F. Riesz's theorem there is a uniquely determined element

$$h \in D(S) \qquad \text{such that} \qquad (Sf, g) = (f, h) \qquad \text{for any } f \in D(S),$$

and $\|h\| \leqslant \|g\|$. Putting

$$h = \tilde{S}_1 g,$$

we obtain a linear operator $\tilde{S}_1$ in H which coincides with S_1 in D(S) because of the equality (2), and moreover $\|\tilde{S}_1 g\| \leqslant \|g\|$, which gives

$$\|\tilde{S}_1\| \leqslant 1. \tag{3}$$

We observe that, by construction, $R(\tilde{S}_1) \subseteq D(S)$, and that, for $f \in D(S)$ and $g \subset H$,

$$(Sf, g) = (f, \tilde{S}_1 g). \tag{2'}$$

We shall now extend the operator S_2 on to H in such a way that the inequality

$$\|\tilde{S}_1 f\|^2 + \|\tilde{S}_2 f\|^2 \leqslant \|f\|^2, \tag{4}$$

which follows from (1) when $f \in D(S)$, shall hold for any $f \in H$. To do this, we define in H a bilinear functional $[f, g]$ by the formula

$$[f, g] = (f, g) - (\tilde{S}_1 f, \tilde{S}_1 g). \tag{5}$$

Moreover, by (3), we shall have

$$[f, f] = (f, f) - (\tilde{S}_1 f, \tilde{S}_1 f) \geqslant 0,$$

but here equality to zero when $f \neq 0$ is not excluded. So the bilinear functional (5) is a quasi-scalar product, as defined in §3. The linear manifold N of all those $f \in \mathrm{H}$ for which $[f, f] = 0$ is closed, because the operator $\tilde{S}_1$ is continuous. By passing from H to the factor space we obtain a linear metrized system H/N. If this system is not complete, then we can complete it in the usual way (see §4), and as a result we obtain a new Hilbert space **H**.

It follows from the inequality (4), which can now be re-written as

$$\|\tilde{S}_2 f\|^2 \leqslant [f, f], \tag{6}$$

that, for $\varphi, \psi \in \mathrm{D}(S)$ and $\varphi - \psi \in \mathrm{N}$, we have $S_2\varphi = S_2\psi$. This means that we can regard S_2 as an operator from **H** into H. The domain of this operator is the linear manifold

$$\mathbf{F} = \mathrm{D}(S)/\mathrm{N} \subset \mathbf{H},$$

and its range lies in $\mathrm{H} \ominus \mathrm{D}(S)$.

Further, it follows from (6) that we can define by continuity in **H** the operator S_2 on the closure $\bar{\mathbf{F}}$ of the manifold **F** in **H**. But when this has been done, it is clear that the values of the operator S_2 will, as before, lie in $\mathrm{H} \ominus \mathrm{D}(S)$, and the inequality (6) will still hold.

Now let $\boldsymbol{P}$ denote the ortho-projector in **H** on to **F**. We extend S_2 on to **H** by putting

$$\tilde{S}_2 \boldsymbol{f} = S_2 \boldsymbol{P} \boldsymbol{f} \qquad (\boldsymbol{f} \in \mathbf{H}).$$

Clearly the operator $\tilde{S}_2$ maps the whole of **H** on to $\mathrm{H} \ominus \mathrm{D}(S)$, and, by (5), for any $\boldsymbol{f} \in \mathbf{H}$,

$$\|\tilde{S}_2 \boldsymbol{f}\|^2 = \|S_2 \boldsymbol{P}\boldsymbol{f}\|^2 \leqslant [\boldsymbol{P}\boldsymbol{f}, \boldsymbol{P}\boldsymbol{f}] \leqslant [\boldsymbol{f}, \boldsymbol{f}]. \tag{7}$$

The constructed extension $\tilde{S}_2$ can be regarded as an operator in the original space if we put, for any $f \in \mathrm{H}$,

$$\tilde{S}_2 f = \tilde{S}_2 \boldsymbol{f},$$

where $\boldsymbol{f}$ is that element of **H** to which the sub-set containing f in H corresponds. Moreover, by (5), we shall have

$$\|\tilde{S}_2 f\|^2 \leqslant [\boldsymbol{f}, \boldsymbol{f}] = (f, f) - (\tilde{S}_1 f, \tilde{S}_1 f),$$

i.e., the inequality (4) will hold for any $f \in \mathrm{H}$.

We now construct an operator T according to the formula

$$Tf = \tilde{S}_1 f + \tilde{S}_2 f \qquad (f \in \mathrm{H}). \tag{8}$$

Since the terms on the right in (8) are orthogonal, we find, using (4), that

$$\|Tf\|^2 = \|\tilde{S}_1 f\|^2 + \|\tilde{S}_2 f\|^2 \leqslant \|f\|^2.$$

Also, for any $\varphi \in \mathrm{D}(S)$, we have

$$T\varphi = \tilde{S}_1 \varphi + \tilde{S}_2 \varphi = S_1 \varphi + S_2 \varphi = S\varphi.$$

Thus the operator T is an extension (in general, not self-adjoint) of the operator S on to the whole of H, preserving the norm of S. We shall show that the operator T^* has the same properties.

For any $\varphi \in \mathrm{D}(S)$ and any $g \in \mathrm{H}$, we have, using (2′), that

$$(T^*\varphi, g) = (\varphi, Tg) = (\varphi, \tilde{S}_1 g + \tilde{S}_2 g) = (\varphi, \tilde{S}_1 g) = (S\varphi, g),$$

and so it is indeed true that $T^*\varphi = S\varphi$ for $\varphi \in \mathrm{D}(S)$, *i.e.*, $T^* \supset S$. But we also have

$$\|T^*\| = \|T\| = 1.$$

Finally, putting

$$\tilde{S} = \tfrac{1}{2}(T + T^*),$$

we obtain the required self-adjoint extension $\tilde{S}$ of the operator S, which preserves the norm. Q.E.D.

To describe *all* the self-adjoint extensions of the operator S which preserve the norm we need the following

Lemma. *Let B be a bounded, positive operator in* H, *and let* G *be a sub-space of* H. *Let $\mathscr{C}$ denote the set of all self-adjoint operators C such that*

$$C \leqslant B, \tag{9}$$

$$Cg = 0 \qquad (g \in \mathrm{G}). \tag{10}$$

Then there is in $\mathscr{C}$ a maximal operator,
i.e., an operator which is $\geqslant$ any other operator in the set $\mathscr{C}$,
and this maximal operator $\bar{C}$ can be expressed in the form

$$\bar{C} = B^{1/2} Q B^{1/2}, \tag{11}$$

where Q is the operator of orthogonal projection on to H⊖F,
and F *is the manifold of all vectors $B^{1/2}g$* $(g \in \mathrm{G})$.

Proof. (The proof given here is due to I. M. Gelfand; see Krein [60].) If $C \in \mathscr{C}$, then, by (10), for any $h \in \mathrm{H}$ and any $g \in \mathrm{G}$, we have

$$(Ch, g) = 0,$$

and therefore

$$(Ch, h) = (C[h-g], h-g).$$

Hence, by (9) and (10),

$$(Ch, h) \leqslant (B[h-g], h-g) = \|B^{1/2}h - B^{1/2}g\|^2,$$

and therefore

$$(Ch, h) \leqslant \inf_{g \in G} \|B^{1/2}h - B^{1/2}g\|^2.$$

This inequality implies that

$$(Ch, h) \leqslant \|QB^{1/2}h\|^2 = (B^{1/2}QB^{1/2}h, h) = (\bar{C}h, h) \qquad (h \in H)$$

as was to be proved. Hence the lemma.

Now let $\tilde{S}_0$ be a fixed, and $\tilde{S}$ be an arbitrary, self-adjoint extension of the operator S, both preserving its norm ($\|S\| = 1$). Then the operator C defined by

$$\tilde{S} = \tilde{S}_0 + C \tag{12}$$

has the following two properties

$$-(I + \tilde{S}_0) \leqslant C \leqslant (I - \tilde{S}_0) \tag{13}$$

$$Cf = 0 \qquad \text{for} \qquad f \in \mathrm{D}(S). \tag{14}$$

It is also clear that, conversely, if a self-adjoint operator C has the properties (13) and (14), then the operator $\tilde{S}$ defined by (12) will be a self-adjoint extension of the operator S, preserving its norm.

Hence, using the lemma, it is easy to obtain a description of all self-adjoint extensions of an operator S which preserve its norm.

To do this, we let Q_1 and Q_2 denote the ortho-projectors on to the orthogonal complements of the manifolds $(I + \tilde{S}_0)^{1/2}\mathrm{D}(S)$ and $(I - \tilde{S}_0)^{1/2}\mathrm{D}(S)$ respectively. Further, we construct the operators

$$\bar{C}_1 = (I + \tilde{S}_0)^{1/2} Q_1 (I + \tilde{S}_0)^{1/2} \tag{15}$$

and

$$\bar{C}_2 = (I - \tilde{S}_0)^{1/2} Q_2 (I - \tilde{S}_0)^{1/2}. \tag{16}$$

Finally we define two self-adjoint operators S' and S'' by putting

$$S' = \tilde{S}_0 - \bar{C}_1, \qquad S'' = \tilde{S}_0 + \bar{C}_2. \tag{17}$$

Theorem 2. *A self-adjoint operator T is a norm-preserving extension of the operator S if and only if it satisfies the double inequality*

$$S' \leqslant T \leqslant S''. \tag{18}$$

Proof. If T is a self-adjoint, norm-preserving extension of the operator S, then the condition (18) follows immediately from (13), (14) and the lemma.

Conversely, suppose that T satisfies (18). If T *is* an extension of the operator S, then it follows immediately from (13), (14) and the lemma that $\|T\| = \|S\|$. So it only remains to show that any operator which satisfies the double inequality (18) is an extension of the operator S.

Putting

$$W = T - S'$$

we obtain from (18) the double inequality

$$0 \leqslant W \leqslant S'' - S',$$

and since $S' \supset S$ and $S'' \supset S$, we have, for any $f \in \mathrm{D}(S)$,

$$0 \leqslant (Wf, f) \leqslant ([S'' - S']f, f) = 0,$$

and so

$$(Wf, f) = 0 \qquad (f \in \mathrm{D}(S)).$$

From this equality and the fact that the operator W is non-negative it follows that $Wf = 0$ for $f \in \mathrm{D}(S)$, *i.e.*, $Tf = S'f$ for $f \in \mathrm{D}(S)$, and therefore $T = \tilde{S}$. Q.E.D.

With each norm-preserving, self-adjoint extension $\tilde{S}$ of the operator S ($\|S\| = 1$) we shall associate an '$\tilde{S}$-scalar product' and an '$\tilde{S}$-norm' by putting

$$(f, g)_{\tilde{S}} = (f, g) + (\tilde{S}f, g)$$

and $\qquad (f, g \in \mathrm{H})$

$$\|f\|_{\tilde{S}} = \sqrt{((f, f)_{\tilde{S}})}.$$

Theorem 3. *Let T be a norm-preserving, self-adjoint extension of an operator S: $\|T\| = \|S\| = 1$. Then T will coincide with the lower bound S' in the formula* (18) *if and only if the manifold* $\mathrm{D}(S)$ *is dense in* H *relative to the T-norm.*

Proof. By taking T as the operator $\tilde{S}_0$ in constructing the formulae (17), we see that the equality $T \equiv \tilde{S}_0 = S'$ is equivalent to the equality $\bar{C}_1 = 0$. On the other hand, by (15),

$$(\bar{C}_1 f, f) = \|Q_1(I + \tilde{S}_0)^{1/2} f\|^2 \qquad (f \in \mathrm{H}).$$

By the definition of Q_1, the right-hand side is

$$\inf_{g \in \mathrm{D}(S)} \|(I + \tilde{S}_0)^{1/2} f - (I + \tilde{S}_0)^{1/2} g\| = \inf_{g \in \mathrm{D}(S)} \|f - g\|^2_{\tilde{S}_0}.$$

Thus

$$(\bar{C}_1 f, f) = \inf_{g \in D(S)} \|f - g\|^2_{\tilde{S}_0} \qquad (f \in H).$$

from which the theorem follows.

In conclusion we observe that, although some extension $\tilde{S}_0$ has to be given in order to construct the operators S' and S'' in accordance with the formulae (15)–(17), in actual fact, as follows from Theorem 2 itself, the operators S' and S'' do not depend on the choice of $\tilde{S}_0$ but are the unique extremal elements (*i.e.*, the least and greatest elements) of the set of all self-adjoint extensions of the operator S which preserve its norm. We shall therefore put

$$S' = \tilde{S}_{\min}, \qquad S'' = \tilde{S}_{\max},$$

and in future write instead of (18)

$$\tilde{S}_{\min} \leqslant T \leqslant \tilde{S}_{\max}. \tag{19}$$

§109. Self-adjoint extensions of a semi-bounded symmetric operator which preserve its lower bound

As already mentioned in §107, the results obtained in §108 enable us to prove Theorem 3 in §107 if we make use of the transformation (2) in §107. We begin this section by studying this bilinear transformation.

Let A be a closed, symmetric, positive operator with its domain $D(A)$ dense in H. Since $\lambda = -1$ is a point of regular type for the operator A, the manifold $(I + A)D(A)$ is a sub-space, and, for any $g \in (I + A)D(A)$, the equation $Af + f = g$ has a unique solution $f \in D(A)$. So we can define an operator S on the sub-space $D(S) = (I + A)D(A)$ by the formulae

$$g = f + Af \tag{1}$$

$$Sg = f - Af \qquad (f \in D(A)). \tag{2}$$

It is easily checked that this operator S is symmetric (with its domain, in general, not dense), and that its norm does not exceed 1. For, we have, for any $g_1 = f_1 + Af_1 \in D(S)$ and $g_2 = f_2 + Af_2 \in D(S)$,

$$(Sg_1, g_2) = (f_1 - Af_1, f_2 + Af_2) = (f_1, f_2) - (Af_1, Af_2),$$

$$(g_1, Sg_2) = (f_1 + Af_1, f_2 - Af_2) = (f_1, f_2) - (Af_1, Af_2).$$

Also, for any $g = f + Af \in D(S)$,

$$\|Sg\|^2 = (f - Af, f - Af) = \|f\|^2 - 2(Af, f) + \|Af\|^2$$

$$\leqslant \|f\|^2 + 2(Af, f) + \|Af\|^2 = (f + Af, f + Af) = \|g\|^2,$$

and so

$$\|S\| \leqslant 1. \tag{3}$$

The operator S has another special property in addition to the above, namely, the number $\mu = -1$ is not an eigenvalue. For,

$$Sg + g \neq 0 \qquad \text{(for } g \neq 0\text{)}, \tag{4}$$

because, if $g \neq 0$, it follows from (1) that $f \neq 0$, and then (1) and (2) imply that

$$Sg + g = 2f \neq 0.$$

(Notice that, in deriving these properties of the operator S, we have not used the denseness of $D(A)$ in H.)

Now suppose, conversely, that S is a symmetric operator in H, defined on a sub-space $D(S) \subset H$, and having the properties (3) and (4).

By virtue of (4) we can define on the linear manifold

$$D(A) = (I + S)D(S)$$

an operator A by the formulae

$$f = \tfrac{1}{2}(g + Sg) \tag{5}$$

$$Af = \tfrac{1}{2}(g - Sg) \qquad (g \in D(S)) \tag{6}$$

It is easily checked that this operator A is symmetric and positive. For, we have, for any $f_1 = \frac{1}{2}(g_1 + Sg_1) \in D(A)$ and $f_2 = \frac{1}{2}(g_2 + Sg_2) \in D(A)$,

$$4(Af_1, f_2) = (g_1 - Sg_1, g_2 + Sg_2) = (g_1, g_2) - (Sg_1, Sg_2),$$

$$4(f_1, Af_2) = (g_1 + Sg_1, g_2 - Sg_2) = (g_1, g_2) - (Sg_1, Sg_2).$$

Also, for any $f = \frac{1}{2}(g + Sg) \in D(A)$,

$$4(Af, f) = (g - Sg, g + Sg) = \|g\|^2 - \|Sg\|^2 \geqslant 0.$$

We observe that the domain $D(A)$ of the operator A may be non-dense in H.

It is clear that the formulae (1) and (2) imply the formulae (5) and (6), and conversely. The transformation defined by these formulae can also be expressed in the form

$$S = (I - A)(I + A)^{-1} \tag{7}$$

or $$A = (I - S)(I + S)^{-1}. \tag{8}$$

It is easily checked that, if either of the operators (7) or (8) is self-adjoint, then so is the other.

As already mentioned at the end of §107, to prove Theorem 3 of §107 it suffices to establish the following proposition.

Theorem 1. Any positive, symmetric operator A *whose domain is dense in* H *has at least one positive, self-adjoint extension.*

Proof. We let S denote the operator defined by the formula (7), and, using Theorem 1 of §108, we construct its self-adjoint extension $\tilde{S}$ with norm $\|\tilde{S}\| = \|S\|$ ($\leqslant 1$).

It is easily seen that the operator $\tilde{S}$ satisfies the condition (4), *i.e.*,

$$\tilde{S}g + g \neq 0 \quad \text{if} \quad g \neq 0.$$

For, if $\tilde{S}g + g = 0$ held for some $g \neq 0$, then we should also have

$$(g, h + \tilde{S}h) = (g + \tilde{S}g, h) = 0 \qquad \text{for any } h \in \mathrm{H},$$

and, in particular, $\quad (g, h + Sh) = 0 \qquad \text{for any } h \in \mathrm{D}(S);$

but $\quad (I + S)\mathrm{D}(S) = \mathrm{D}(A),$

and if the last scalar product $(g, h + Sh)$ were equal to zero, the domain $\mathrm{D}(A)$ would not be dense in H.

Since the operator $\tilde{S}$ satisfies the conditions (3) and (4), the operator

$$\tilde{A} = (I - \tilde{S})(I + \tilde{S})^{-1}$$

exists, and it, clearly, is a positive extension of the operator A. Since, finally, the operator $\tilde{S}$ is self-adjoint, the operator $\tilde{A}$ also is self-adjoint.

Q.E.D.

Let $\mathscr{A}(A)$ denote the set of all positive, self-adjoint extensions of the operator A, and let $\mathscr{S}(S)$ denote the set of all self-adjoint extensions, of the operator S, related to A by (7), which have a norm $\leqslant 1$. It follows from the proof of Theorem 1 that in the formula

$$\tilde{A} = (I - \tilde{S})(I + \tilde{S})^{-1},$$

where $\quad \tilde{S} \in \mathscr{S}(S), \quad$ and $\quad S = (I - A)(I + A)^{-1},$

the operator $\tilde{A}$ runs over the whole of $\mathscr{A}(A)$ if the operator $\tilde{S}$ runs over the whole of $\mathscr{S}(S)$.

It follows from Theorem 2 of §108 that the general form of the operator $\tilde{A} \in \mathscr{A}(A)$ is given by the formula

$$\tilde{A} = (I - T)(I + T)^{-1},$$

where T is any self-adjoint operator which satisfies the double inequality (19) of §108. The operators

$$A_\mu = (I - \tilde{S}_{\min})(I + \tilde{S}_{\min})^{-1}$$

and

$$A_M = (I - \tilde{S}_{\max})(I + \tilde{S}_{\max})^{-1}$$

correspond to the extreme members $\tilde{S}_{\min}$ and $\tilde{S}_{\max}$ of this double inequality (19).

The operator A has a unique, positive, self-adjoint extension if and only if $A_\mu = A_M$.

In the general case, $A_\mu \neq A_M$. In this case, both the operators A_μ and A_M have certain extremal properties. Of the two, the greatest interest attaches to the operator A_μ, which is called the *strict extension* of the positive operator A. It was this extension which was obtained by K. Friedrichs in proving the truth of von Neumann's conjecture. We shall devote the rest of this section to an examination of this strict self-adjoint extension A_μ of a positive, symmetric operator A.

We shall define a mode of convergence of elements in the manifold $\mathrm{D}(A)$ as follows. A sequence of elements $f_n \in \mathrm{D}(A)$ will be called *A-convergent* if the sequence $\{f_n\}$ converges in H and if, in addition,

$$\lim_{m,n\to\infty} (A(f_n - f_m), f_n - f_m) = 0. \tag{9}$$

We shall denote the closure of $\mathrm{D}(A)$ in the sense of A-convergence by $\mathrm{D}[A]$.

It is clear that A-convergence is the same as convergence relative to the A-norm if we define the A-norm by the relations

$$(f, g)_A = (f, g) + (Af, g), \qquad \|f\|_A = \sqrt{(f, f)_A}.$$

Lemma. (K. Friedrichs)
If $\{g_n\}_1^\infty$ and $\{f_n\}_1^\infty$ are any two sequences in $\mathrm{D}(A)$ which are A-convergent to $g \in \mathrm{D}[A]$ and $f \in \mathrm{D}[A]$ respectively, then

$$\lim_{n\to\infty} (Ag_n, f_n) \tag{10}$$

exists. Moreover, this limit does not depend on the choice of the sequences $\{g_n\}_1^\infty$, $\{f_n\}_1^\infty$, and, for any $\tilde{A} \in \mathcal{A}(A)$ the following relations hold:

$$\mathrm{D}[A] \subset \mathrm{D}(\tilde{A}^{1/2}),$$

$$\lim_{n\to\infty} (Ag_n, f_n) = (\tilde{A}^{1/2} g, \tilde{A}^{1/2} f) \qquad (g, f \in \mathrm{D}[A]).$$

Proof. From the spectral representation of a self-adjoint, positive operator $\tilde{A}$ the inclusion $D(\tilde{A}) \subset D(\tilde{A}^{1/2})$ easily follows. Therefore the relation (9) can be expressed in the form

$$\lim_{m,n\to\infty} \|\tilde{A}^{1/2}f_n - \tilde{A}^{1/2}f_m\| = 0,$$

whence, because of the convergence of f_n to f and the fact that the operator $\tilde{A}^{1/2}$ is closed, we obtain the inclusion $f \in D(\tilde{A}^{1/2})$ and the relation

$$\lim_{n\to\infty} \tilde{A}^{1/2}f_n = \tilde{A}^{1/2}f.$$

Similarly

$$\lim_{n\to\infty} \tilde{A}^{1/2}g_n = \tilde{A}^{1/2}g.$$

From the relations obtained, the truth of both assertions of the lemma follows at once. Q.E.D.

The following theorem, due to M. G. Krein, gives a characterization of the strict extension A_μ of a positive operator A.

Theorem 2. Among all the possible self-adjoint extensions, semi-bounded below, of a positive, symmetric operator A there is a unique extension such that $D(\tilde{A}) \subset D[A]$. *This extension is the operator* A_μ, *and for it*

$$D[A_\mu] = D[A]. \tag{11}$$

Proof. We restrict our attention initially to extensions $\tilde{A} \in \mathscr{A}(A)$, so that

$$\tilde{A} = (I - \tilde{S})(I + \tilde{S})^{-1},$$

where $\tilde{S} \in \mathscr{S}(S)$, and $S = (I - A)(I + A)^{-1}$.

If $D(\tilde{A}) \subset D[A]$, (12)

then, clearly, the manifold $D(A)$ will be dense, relative to the $\tilde{A}$-norm, in $D(\tilde{A})$.

On the other hand, if $f \in D(\tilde{A})$, then

$$f = \tfrac{1}{2}(g + \tilde{S}g), \qquad \text{where} \qquad g = f + \tilde{A}f,$$

and so

$$2[(f, f) + (\tilde{A}f, f)] = (g, g) + (\tilde{S}g, g),$$

i.e.,

$$2\|f\|_{\tilde{A}} = \|g\|_{\tilde{S}}.$$

Therefore the denseness of D(A) in D($\tilde{A}$) relative to the $\tilde{A}$-norm implies the denseness of the manifold D(S) = (I + A)D(A), relative to the $\tilde{S}$-norm, in D($\tilde{S}$) = (I + $\tilde{A}$)D($\tilde{A}$) = H. But then, by Theorem 3 of §108, we shall have $\tilde{S} = S_{\min}$, and therefore

$$\tilde{A} = A_\mu.$$

Conversely, if $\tilde{A} = A_\mu$, then $\tilde{S} = S_{\min}$, and by Theorem 3 of §108 again, the manifold D(S) will be dense in H relative to the $\tilde{S}$-norm, and consequently the manifold D(A) will be dense in D($\tilde{A}$) relative to the $\tilde{A}$-norm. But since D[A] is the closure of D(A) relative to the $\tilde{A}$-norm, D($\tilde{A}$) ⊂ D[A].

On the other hand, D($\tilde{A}$) ⊃ D(A). Therefore the closure of D($\tilde{A}$) relative to the $\tilde{A}$-norm coincides with D[A], *i.e.*,

$$\mathrm{D}[A] = \mathrm{D}[\tilde{A}],$$

and this is the same as (11).

It remains now to show that (12) implies $\tilde{A} = A_\mu$ not only when $\tilde{A} \in \mathcal{A}(A)$, but also in the more general case when $\tilde{A}$ is an arbitrary self-adjoint extension, semi-bounded below, of the operator A.

To do this for a given $\tilde{A}$, it is sufficient to choose a number $\alpha > 0$ such that $\tilde{A} + \alpha I > 0$. Then from (12) we obtain the relation

$$\mathrm{D}(\tilde{A} + \alpha I) = \mathrm{D}(\tilde{A}) \subset \mathrm{D}[A] = \mathrm{D}[A + \alpha I],$$

and so, from what we have already proved, we have

$$\tilde{A} + \alpha I = (A + \alpha I)_\mu.$$

Since, in particular, (12) holds for $A = A_\mu$, we shall have

$$A_\mu + \alpha I = (A + \alpha I)_\mu,$$

from which it follows that $\tilde{A} = A_\mu$, and the theorem has been proved completely.

9

Generalized extensions and generalized spectral functions of symmetric operators

§110. Generalized resolutions of the identity. M. A. Naimark's theorem†

Every 1-parameter family of operators F_t which has the following properties

(A) $F_{t_2} - F_{t_1}$ is a bounded positive operator if $t_2 > t_1$;
(B) $F_{t-0} = F_t$;
(C) $F_{-\infty} = 0, \quad F_\infty = 1$

will be called a *generalized resolution of the identity.*

In contrast to the definition of an ordinary resolution of the identity (see §67), here it is not required that F_t be a projection; and hence the requirement of orthogonality

$$F_u F_v = F_s \qquad (s = \min\{u, v\}) \tag{1}$$

is also dropped, because (1) and the listed conditions (A), (B), (C) would imply that F_t is necessarily a projection operator.

From Theorem 3 in §33 on monotone sequences of operators (and the remark after that theorem) it follows that whenever an operator function F_t has the property (A) it can always be suitably normalized so that the condition (B) is satisfied.

An ordinary resolution of the identity is a special case of a generalized resolution of the identity. Sometimes a generalized resolution of the identity is simply called a resolution of the identity, and the ordinary resolution of the identity is then called an *orthogonal resolution of the identity.*

In addition to F_t we also introduce a positive, additive, operator function of an interval $\Delta = [t_1, t_2)$ by putting

$$F(\Delta) = F_\Delta = F_{t_2} - F_{t_1}.$$

† See Naimark [86, 87].

A very simple example of a generalized resolution of the identity is given by the operator function F_t defined by

$$F_t = \mu_1 E_t^{(1)} + \mu_2 E_t^{(2)},$$

where $E_t^{(1)}, E_t^{(2)}$ are any two orthogonal resolutions of the identity, and μ_1, μ_2 are positive numbers such that $\mu_1 + \mu_2 = 1$.

A more instructive example is the operator function F_t defined as follows.

Let E_t be an orthogonal resolution of the identity for the space H, let G be some sub-space of H, and let P be the operator of orthogonal projection on to G. We put

$$F_t = PE_t$$

and regard F_t as an operator operating in the space G. It is then easy to see that the operator function F_t satisfies the conditions (A), (B), (C), and that it is therefore a resolution of the identity —— in general, not an orthogonal one —— for the space G.

M. A. Naimark proved that any generalized resolution of the identity for a space H can be constructed, by the method just described, by first embedding the space H in some space H^+.

To prove Naimark's theorem we need a certain important method of constructing Hilbert spaces, and we begin by describing this method. For this purpose we introduce the concept of an Hermitian-positive function.

A numerical function $\Phi(\mathfrak{f}, \mathfrak{g})$ defined on any pair of elements $\mathfrak{f}, \mathfrak{g}$ of some set $\mathfrak{H}$ is said to be *Hermitian-positive* if, for any $\mathfrak{f}, \mathfrak{g} \in \mathfrak{H}$,

$$\Phi(\mathfrak{f}, \mathfrak{g}) = \overline{\Phi(\mathfrak{g}, \mathfrak{f})}$$

and if, for any $n < \infty$ and any $\mathfrak{f}_1, \mathfrak{f}_2, \ldots, \mathfrak{f}_n \in \mathfrak{H}$, the Hermitian form

$$\sum_{i,k=1}^{n} \Phi(\mathfrak{f}_i, \mathfrak{f}_k)\xi_i\bar{\xi}_k$$

is non-negative.

An example of an Hermitian-positive function of pairs of vectors of a Hilbert space is the scalar product, since

$$(f, g) = \overline{(g, f)}$$

and

$$\sum_{i,k=1}^{n} (f_i, f_k)\xi_i\bar{\xi}_k = \left\| \sum_{i=1}^{n} \xi_i f_i \right\|^2 \geqslant 0.$$

Conversely, if we are given an Hermitian-positive function of a pair of elements of an arbitrary set $\mathfrak{H}$, then this set can be embedded in a certain

Hilbert space H^+ such that the scalar product of the elements $\mathfrak{f}, \mathfrak{g} \in \mathfrak{H}$ is defined by the formula

$$(\mathfrak{f}, \mathfrak{g}) = \Phi(\mathfrak{f}, \mathfrak{g}).$$

To construct H^+, we first complete $\mathfrak{H}$ into a linear system $\hat{\mathfrak{H}}$ by introducing formally the finite sums

$$\hat{\mathfrak{f}} = \sum_{i=1}^{n} \xi_i \mathfrak{f}_i$$

for any $n < \infty$ and for arbitrary $\mathfrak{f}_i \in \mathfrak{H}$ and arbitrary numbers $\xi_i \quad (i = 1, \ldots, n)$.

Further, we define in $\hat{\mathfrak{H}}$ the quasi-scalar product (see §3) by putting

$$(\hat{\mathfrak{f}}, \hat{\mathfrak{g}}) = \sum_{i,k=1}^{n} \Phi(\mathfrak{f}_i, \mathfrak{g}_k) \xi_i \bar{\xi}_k,$$

where

$$\hat{\mathfrak{f}} = \sum_{i=1}^{p} \xi_i \mathfrak{f}_i, \qquad \hat{\mathfrak{g}} = \sum_{k=1}^{q} \eta_k \mathfrak{g}_k, \qquad n = \min\{p, q\}.$$

Let $\hat{\mathfrak{N}}$ denote the linear manifold of all those $\hat{\mathfrak{f}} \in \hat{\mathfrak{H}}$ such that $(\hat{\mathfrak{f}}, \hat{\mathfrak{f}}) = 0$, and we then go over to the factor-space

$$R^+ = \hat{\mathfrak{H}} / \hat{\mathfrak{N}}.$$

By completing the linear metrized system R^+ we obtain a certain Hilbert space H^+, which is the one required.

We can now deal with Naimark's theorem.

Theorem. *If F_t is a generalized resolution of the identity for the space* H, *then there is a Hilbert space H^+ which contains* H *as a sub-space, and there is an orthogonal resolution of the identity E_t^+ for the space H^+ such that, for any $f \in H$,*

$$F_t f = P^+ E_t^+ f,$$

where P^+ is the operator of orthogonal projection on to H.

Proof. Let $\mathfrak{H}$ be the set of pairs $\mathfrak{p}$ of the form

$$\mathfrak{p} = \{\Delta, f\},$$

where $\Delta = [t_1, t_2)$ is an arbitrary left-closed sub-interval of the interval $\Omega = [-\infty, \infty)$, and f is an arbitrary vector in H.

Further, we define a function $\Phi(\mathfrak{p}_1, \mathfrak{p}_2)$ on $\mathfrak{H}$ by putting

$$\Phi(\mathfrak{p}_1, \mathfrak{p}_2) = (F_{\Delta_1 . \Delta_2} f_1, f_2)$$

for $\mathfrak{p}_1=\{\Delta_1, f_1\}$, $\mathfrak{p}_2=\{\Delta_2, f_2\}$. Here $\Delta_1 \,.\, \Delta_2$ denotes the intersection of Δ_1 and Δ_2; and, later, $\Delta_1+\Delta_2$ will denote the union.

We shall show that $\Phi(\mathfrak{p}_1, \mathfrak{p}_2)$ is an Hermitian-positive function.

For,

$$\Phi(\mathfrak{p}_1, \mathfrak{p}_2)=(F_{\Delta_1 . \Delta_2}f_1, f_2)=(f_1, F_{\Delta_1 . \Delta_2}f_2)=\overline{(F_{\Delta_1 . \Delta_2}f_2, f_1)}=\overline{\Phi(\mathfrak{p}_2, \mathfrak{p}_1)};$$

but on the other hand

$$\sum_{i,k=1}^{n} \Phi(\mathfrak{p}_i, \mathfrak{p}_k)\xi_i\bar{\xi}_k = \sum_{i,k=1}^{n} (F_{\Delta_i . \Delta_k}f_i, f_k)\xi_i\bar{\xi}_k. \tag{2}$$

If the intervals Δ_i $(i=1, 2, \ldots, n)$ are pairwise disjoint, then

$$\sum_{i,k=1}^{n} (F_{\Delta_i . \Delta_k}f_i, f_k)=\sum_{i=1}^{n} (F_{\Delta_i}f_i, f_i)\,|\xi_i|^2 \geqslant 0. \tag{3}$$

But if the intervals Δ_i are pairwise disjoint only for $i=2, 3, \ldots, n$ and the intervals Δ_1 and Δ_2 coincide, then the sum on the right-hand side of (2) splits into two parts; the part containing the subscripts $3, \ldots, n$ will have the same form as (3), but the part containing the subscripts 1 and 2 will be

$$\sum_{i.k=1}^{2} (F_{\Delta_i . \Delta_k}f_i, f_k)\xi_i\bar{\xi}_k = \sum_{i,k=1}^{2} (F_{\Delta_1}f_i, f_k)\xi_i\bar{\xi}_k = \left(F_{\Delta_1} \sum_{i=1}^{2} \xi_i f_i, \sum_{k=1}^{2} \xi_k f_k\right) \geqslant 0.$$

The case where the intervals Δ_i $(i=1, 2, \ldots, n)$ are in general position reduces to the cases already considered if we introduce additional points of partition so that the given system becomes one in which the intervals either coincide or are pairwise disjoint, and if we use the fact that, if $\Delta_1 \,.\, \Delta_2 - 0$, then

$$(F_{(\Delta_1+\Delta_2) . \Delta_3}f, g)=(F_{(\Delta_1 . \Delta_3+\Delta_2 . \Delta_3)}f, g)=(F_{\Delta_1 . \Delta_3}f, g)+(F_{\Delta_2 . \Delta_3}f, g).$$

Thus the function $\Phi(\mathfrak{p}_1, \mathfrak{p}_2)$ is Hermitian-positive on $\mathfrak{H}$.

Using the method already described, we shall embed $\mathfrak{H}$ in a Hilbert space H^+.

Since we do not want to introduce a new notation for those element $\mathfrak{P}$ of the space H^+ which are sub-sets in $\hat{\mathfrak{H}}$, *i.e.*, for the elements in R^+, let us agree to the following: if some element $\mathfrak{p}$ in $\hat{\mathfrak{H}}$ belongs to $\mathfrak{P}$, then instead of $\mathfrak{P}$ we shall write $\mathfrak{p}$.

We shall add the subscript + to distinguish the scalar product in H^+, and so we have

$$(\mathfrak{p}_1, \mathfrak{p}_2)_+ = \Phi(\mathfrak{p}_1, \mathfrak{p}_2).$$

We now consider the elements in H^+ which are of the form $\{\Omega, \mathfrak{f}\}$. Since

$$(\{\Omega, f\}, \{\Omega, g\})_+ = (F_\Omega f, g)=(f, g),$$

we may identify the pair $\{\Omega, f\}$ with the element f in H. The element $\sum_{k=1}^{n} \xi_k\{\Omega, f_k\}$ of the space H^+ is then identified with the element $\sum_{k=1}^{n} \xi_k f_k$ of the space H. Thus H can be regarded as a sub-space of the space H^+.

We shall next solve the following problem: to find the projection on to the sub-space H of the element $\{\Omega, f\}$ of the space H^+. Denoting the required projection by $\{\Omega, g\}$, we must have

$$(\{\Delta, f\}-\{\Omega, g\}, \{\Omega, h\})_+ = 0 \qquad \text{for any } h \in H,$$

or

$$(\{\Delta, f\}, \{\Omega, h\})_+ - (\{\Omega, g\}, \{\Omega, h\})_+ = (F_\Delta f, h) - (g, h) = (F_\Delta f - g, h) = 0,$$

whence

$$g = F_\Delta f,$$

i.e.,

$$P^+\{\Delta, f\} = \{\Omega, F_\Delta f\}. \tag{4}$$

We now define an operator function E_Δ^+ on any element of the form $\{\Delta', f\} \in H^+$ by putting

$$E_\Delta^+\{\Delta', f\} = \{\Delta . \Delta', f\}. \tag{5}$$

Using the fact that the linear envelope of elements of the form $\{\Delta', f\}$ is dense in H^+, we shall extend the operator E_Δ^+ first by linearity and then by continuity on to the whole of H^+.

The theorem will be proved if we show that the operator function E_Δ^+ is an orthogonal resolution of the identity for the space H^+, since then (4) can be presented in the form

$$P^+ E_\Delta^+ f = P^+ E_\Delta^+\{\Omega, f\} = P^+\{\Delta . \Omega, f\} = P^+\{\Delta, f\} = \{\Omega, F_\Delta f\} = F_\Delta f$$

for any $f \in H$.

Clearly, E_Δ^+ is an additive operator interval-function. Further, it follows from the two equalities

$$(E_\Delta^+)^2\{\Delta', f\} = E_\Delta^+\{\Delta . \Delta', f\} = \{\Delta . \Delta . \Delta', f\} = E_\Delta^+\{\Delta', f\}$$

and

$$(E_\Delta^+\{\Delta', f\}, \{\Delta'', g\})_+ = (\{\Delta . \Delta', f\}, \{\Delta'', g\})_+ = (F_{\Delta . \Delta' . \Delta''} f, g)$$

$$= (\{\Delta', f\}, E_\Delta^+\{\Delta'', g\})_+$$

that E_Δ^+ is a projection operator.

Finally, it is clear that $E_\Omega^+\{\Delta', f\} = \{\Delta', f\}$.

In view of the properties (B) and (C) of the family F_t, we see that E_Δ^+ is an orthogonal resolution of the identity for the space H^+, and the theorem has been proved.

§111. Self-adjoint extensions with emergence from the space. Spectral functions of symmetric operators†

Our further considerations are based on a generalization of the concept of symmetric extensions.

Let A be a symmetric operator operating in a Hilbert space H, and let H^+ be a Hilbert space containing H.

Supplementing the definition in §46, we shall call any symmetric (and in particular, any self-adjoint) operator B^+ which operates in H^+ and which is an extension of the operator A a *symmetric* (and in particular, a *self-adjoint*) *extension* of A. (We assume the domain $D(A)$ to be dense in H, and the domain $D(B^+)$ to be dense in H^+.)

The concept of an *isometric extension* (and in particular, of a *unitary extension*) is generalized in a similar way.

Clearly, the Cayley transform of a symmetric (or self-adjoint) extension B^+ of an operator A is an isometric (or unitary) extension of the Cayley transform of the operator A.

Conversely, the Cayley transform of an isometric (or unitary) extension of an isometric operator is a symmetric (or self-adjoint) extension of the Cayley transform of this operator.

When $H^+ = H$ we obtain merely the usual symmetric and isometric extensions which were previously considered.

Let B^+ be an arbitrary symmetric extension of an operator A. Since the inclusions

$$D(A) \subseteq D(B^+) \cap H \subseteq D(B^+)$$

hold, it is convenient to classify symmetric extensions B^+ of the operator A in the following way:

1. Extensions of the first kind: when $D(A) \neq D(B^+) \cap H = D(B^+)$; these extensions are the same as the usual extensions, because in this case $H^+ = H$.
2. Extensions of the second kind: when $D(A) = D(B^+) \cap H \neq D(B^+)$.
3. Extensions of the third kind: when $D(A) \neq D(B^+) \cap H \neq D(B^+)$.

† See M. A. Naimark [85, 86]. Before Naimark developed the general theory, which we expound in this and the next section, particular results were obtained by A. I. Plesner and by N. I. Akhiezer; see Plesner [99] and Akhiezer [1].

Thus, the ordinary symmetric extensions (*i.e.*, extensions without emergence from the original space) are the same as extensions of the first kind, but symmetric extensions with emergence from the original space may be extensions of the second or third kinds.

It is clear that a maximal operator can have only a symmetric extension of the second kind; for, if B^+ were an extension of the third kind, then P^+B^+ would be, in H, a non-trivial symmetric extension of a maximal operator, and this is impossible.

If a symmetric extension B^+ of an operator A is reduced by some sub-space $G^+ \subset H^+ \ominus H$, let us agree to exclude the sub-space G^+ from H^+ (*i.e.*, we shall replace the space H^+ by the space $H^+ \ominus G^+$, and replace the operator B^+ by its restriction to $H^+ \ominus G^+$).

As a result of this convention, a self-adjoint operator does not admit of any symmetric extensions.

We can now direct our attention to the fundamental theorem of this section.

Theorem 1. *A symmetric operator* A, *with an arbitrary deficiency index* $(\mathfrak{m}, \mathfrak{n})$, *operating in a Hilbert space* H *can be extended into a symmetric operator* B^+ *operating in a space* $H^+ \supseteq H$.

Proof. We construct a symmetric operator A', which has the deficiency index $(\mathfrak{n}, \mathfrak{m})$, in some space H' of sufficiently large dimension; (we could, for example, take H' to be isomorphic to H, and A' to be isomorphic to $(-1)A$).

We now construct the space

$$H^+ = H \oplus H'$$

and in this space H^+ we introduce the symmetric operator

$$A^+ = A \oplus A'.$$

Clearly, the operator A^+ is a symmetric extension of the second kind of the operator A, and to prove the theorem it suffices to establish that the operator A' can be extended into a self-adjoint operator; to do this we have to verify that the operator A^+ has equal deficiency numbers. If $\mathcal{I}(z) \neq 0$ we obtain

$$\begin{aligned} M_z^+ &= (A^+ - zI^+)D(A^+) = (A^+ - zI^+)(D(A) \oplus D(A')) \\ &= (A - zI)D(A) \oplus (A' - zI')D(A') = M_z \oplus M_z', \end{aligned}$$

or changing over to the orthogonal complements of the manifolds M_z^+, M_z, M_z' in the spaces H^+, H, and H' respectively, we have

$$N_z^+ = N_z \oplus N_z',$$

from which it follows that the deficiency index of the operator A' is $(\mathfrak{m}+\mathfrak{n}, \mathfrak{m}+\mathfrak{n})$. Q.E.D.

By choosing the space H' and the operator A' in various ways, we shall obtain various self-adjoint extensions of the operator A.

The operator B^+ constructed in the proof of Theorem 1 will, in general, be an extension of the third kind of the original symmetric operator A. However, it is always possible to arrange to obtain an operator B^+ which is an extension of the second kind of the operator A. To do this it suffices to construct a unitary operator U^+ such that

$$U^+ \mathrm{N}_z = \mathrm{N}'_{\bar{z}},$$

and then the required operator B^+ will be the Cayley transform of U^+.

Using Theorem 1 we now show that arbitrary symmetric operators have integral representations similar to the spectral resolutions of self-adjoint operators, which we studied in Chapter 6.

Suppose, then, that a symmetric operator A in H is given. We extend it into a self-adjoint operator B^+, passing, if necessary, from the original space H to a space H^+. (We do not assume here that this extension is carried out by the special method which we adopted in the proof of Theorem 1.) Let $E^+(\Delta)$ be the resolution of the identity for the operator B^+, and let P^+ be the operator projecting H^+ on to H. Finally, let

$$F(\Delta) = P^+ E^+(\Delta).$$

Then for any two elements f, g with $f \in \mathrm{D}(B^+)$, $g \in \mathrm{H}^+$, we have

$$(B^+ f, g) = \int_{-\infty}^{\infty} t\, d(E_t^+ f, g),$$

$$\|B^+ f\|^2 = \int_{-\infty}^{\infty} t^2\, d(E_t^+ f, f).$$

If, in particular, $f \in \mathrm{D}(A)$ and $g \in \mathrm{H}$, then these formulae can be written in the form

$$(Af, g) = \int_{-\infty}^{\infty} t\, d(F_t f, g), \tag{1}$$

$$\|Af\|^2 = \int_{-\infty}^{\infty} t^2\, d(F_t f, f). \tag{2}$$

We have thus obtained an integral representation of an arbitrary symmetric operator which brings to mind the spectral resolution of a self-adjoint operator.

On the basis of this similarity, we adopt the following definition, which generalizes the concept of the spectral function of a self-adjoint operator.

Definition. If A is a symmetric operator, and if F_t is a generalized resolution of the identity such that the formulae (1) and (2) hold for any $f \in D(A)$ and any $g \in H$, then F_t is called a *spectral function* for the operator A.

Before comparing the formulae (1) and (2) with the spectral resolution of a self-adjoint operator obtained earlier, we show that the method by which we arrived at these formulae is general; that is, the following theorem holds.

Theorem 2. Every spectral function of a symmetric operator A operating in a space H has the form

$$F_t = P^+ E_t^+ ,$$

where E_t^+ is a resolution of the identity for some self-adjoint extension B^+ of the operator A, which is obtained by emerging from the space H into a space $H^+ \supseteq H$, and P^+ is the operator which projects H^+ on to H.

Proof. We construct the space H^+ in accordance with Naimark's theorem, and in it an orthogonal resolution of the identity E_t^+ such that

$$F_t = P^+ E_t^+ . \tag{3}$$

We shall show that the operator B^+ defined by

$$B^+ f = \int_{-\infty}^{\infty} t \, dE_t^+ f$$

on elements $f \in H^+$ which satisfy the condition

$$\int_{-\infty}^{\infty} t^2 \, d(E_t^+ f, f) < \infty$$

is a self-adjoint extension of the operator A.

The self-adjointness of the operator B^+ follows from the theorem in §72.

If $f \in D(A)$, then $f \in D(B^+)$, because

$$\int_{-\infty}^{\infty} t^2 \, d(E_t^+ f, f) = \int_{-\infty}^{\infty} t^2 \, d(F_t f, f) = \|Af\|^2 < \infty.$$

Further, if $f \in D(A)$ and $g \in H$, then the relations

$$(Af, g) = \int_{-\infty}^{\infty} t \, d(F_t f, g) = \int_{-\infty}^{\infty} t \, d(E_t^+ f, g) = (B^+ f, g)$$

imply that

$$Af = P^+ B^+ f \qquad (f \in D(A)). \tag{4}$$

But if $f \in D(A)$, then

$$\|Af\|^2 = \int_{-\infty}^{\infty} t^2\, d(F_t f, f) = \int_{-\infty}^{\infty} t^2\, d(E_t^+ f, f) = \|B^+ f\|^2. \tag{5}$$

It follows from (4) and (5) that

$$Af = B^+ f \qquad (f \in D(A)).$$

Moreover, it can be taken that the operator B^+ is not reduced by any sub-space in $H^+ \ominus H$.

For, if a sub-space G of $H^+ \ominus H$ reduced the operator B^+, then it would reduce the resolution of the identity E_t^+, and excluding G from H^+ would be equivalent to excluding rhe restriction of E_t^+ to G, and this would not affect the formula (3). Hence the theorem.

If the spectral function F_t for a symmetric operator is expressed in the form (3), where E_t^+ is the resolution of the identity for a self-adjoint extension B^+ of the operator A, then we shall say that the *spectral function* F_t *is generated by the self-adjoint extension* B^+.

Thus, *every self-adjoint extension of the operator A generates a spectral function for this operator, and conversely, each spectral function for an operator A is generated by some one of its self-adjoint extensions.*

We now present a number of facts which show how far the analogy between the representations (1) and (2) and the spectral representation of a self-adjoint operator extends.

1°. The resolution of the identity for a self-adjoint operator——as defined in Chapter 6——is its only spectral function, in the sense of the definition given in this section. A self-adjoint operator has no other spectral function, because, by the above result, every such spectral function has to be generated by some self-adjoint extension, but a self-adjoint operator does not have any self-adjoint extensions.

2°. The formula (2) implies that, for $f \in D(A)$,

$$\int_{-\infty}^{\infty} t^2\, d(F_t f, f) < \infty, \tag{6}$$

but the converse is, in general, false. It can be asserted only that if, for some $f \in H$, the integral (6) converges, then

$$\int_{-\infty}^{\infty} t^2\, d(E_t^+ f, f) < \infty, \tag{7}$$

so that the vector f belongs to the manifold $D(B^+) \cap H$, which coincides with $D(A)$ if and only if B^+ is an extension of the second kind of the operator A.

Thus, *those (and only those) spectral functions of the operator A which are generated by self-adjoint extensions of the second kind of the operator A characterize the domain* D(A) *as the set of vectors for which the integral* (6) *converges.*

In particular, the spectral functions of maximal operators have this property, because such operators admit only extensions of the second kind.

In the general case, the spectral function F_t of an operator A which is generated by a self-adjoint extension B^+ of A characterizes the domain of an operator $C = P^+B^+$, regarded as an operator in H, which is a symmetric extension of the first kind for A with a maximal domain D(C) satisfying the condition

$$D(A) \subset D(C) \subset D(B^+), \tag{8}$$

i.e., with $D(C) = D(B^+) \cap H$.

From this point of view it would be natural to regard the resolution of the identity F_t as the spectral function of the operator C and not of the operator A, and, correspondingly, to include in the definition the requirement that the resolution of the identity F_t, in addition to giving the representations (1) and (2), should also define the domain D(A) by the inequality (6).

However, from the point of view in later considerations it turns out to be inconvenient to include in the definition the extra requirement regarding the characteristics of the domain D(A).

Thus the definition allows one and the same resolution of the identity to be the spectral function of two different operators (for example, of the operators A and C considered above).

In particular, the spectral function of some self-adjoint extension B of the first kind of a given operator A with equal deficiency numbers is, by the definition, the spectral function of the operator A itself and of any operator C which satisfies the condition $C \subseteq B$.

3°. It follows from the above that, corresponding to each symmetric operator there is a set of spectral functions, and that one and the same spectral function corresponds to some set of symmetric operators.

The question then arises whether every given generalized resolution of the identity is the spectral function of some symmetric operator.

The corresponding question for orthogonal resolutions of the identity was answered affirmatively by the theorem in §72.

If, however, F_t is a non-orthogonal resolution of the identity, then it is impossible to say in advance that the set of vectors for which the integral in (6) exists will be dense in H.

Actually, as M. A. Naimark pointed out, there are resolutions of the identity which are such that the integral in (6) does not converge for any non-zero f in H.

As an example of this sort (for which we are indebted to M. A. Krasnoselsky) we consider the operator function defined on the space H by the equalities

$$F_t = \begin{cases} 0, & t \leq 0, \\ e^{-1/t} I, & t > 0. \end{cases}$$

F_t clearly satisfies the conditions (A), (B), (C) in §110, and so it is a resolution of the identity. But this resolution of the identity is not the spectral function of any symmetric operator at all, because the integral

$$\int_{-\infty}^{\infty} t^2 \, d(F_t f, f) = (f, f) \int_0^{\infty} t^2 \, de^{-1/t} = (f, f) \int_0^{\infty} e^{-1/t} \, dt$$

diverges for every non-zero-vector f.

4°. It was explained in Chapter 6 that for self-adjoint operators the representation (1) holds in the strong sense.

In the case of a non-orthogonal resolution of the identity the passage from the weak representation (1) to a strong representation, of the form

$$Af = \int_{-\infty}^{\infty} t \, dF_t f \tag{9}$$

cannot be effected, and we shall regard (9) as being merely a symbolic expression of the formulae (1) and (2).

In general, the concept of the integral in the strong sense is not introduced for non-orthogonal resolutions of the identity.

5°. From the orthogonality property of the spectral function E_t of a self-adjoint operator A it follows that, for any finite interval Δ of the real line, the vector $E(\Delta)f \quad (f \in H)$ belongs to $D(A)$, and

$$(AE(\Delta)f, g) = \int_{\Delta} t \, d(E_t f, g) \qquad \text{for any } g \in H.$$

This important circumstance, which we used repeatedly in Chapter 6, no longer holds for the integral representations, now under consideration, of symmetric operators (even if we restricted our attention to spectral functions generated by extensions of the second kind).

However, the following proposition does hold.

If A is a symmetric operator and F_t is its spectral function, if Δ is a finite interval of the real line, and if g is any vector in H, *then*

$$F(\Delta)g \in D(A^*),$$

and

$$(A^* F(\Delta)g, h) = \int_{\Delta} t \, d(F_t g, h) \qquad \textit{for any } h \in H.$$

For, if B^+ is the extension which generates F_t, and if P^+ is the operator projecting H^+ on to H, then, for any $f \in D(A)$ and any $g \in H$,

$$(Af, F(\Delta)g) = (B^+f, P^+E^+(\Delta)g) = (B^+f, E^+(\Delta)g)$$

$$= \int_\Delta t\, d(E_t^+f, g) = \int_\Delta t\, d(E_t^+f, P^+g) = \int_\Delta t\, d(F_tf, g). \quad (10)$$

But the integral

$$\int_\Delta t\, d(F_th, g)$$

is a bilinear functional of h, g in H, and therefore it has a representation (h, Cg), where C is some bounded operator.

Hence it follows that

$$(Af, F(\Delta)g) = (f, g^*)$$

and therefore

$$F(\Delta)g \in D(A^*).$$

Therefore it follows from (10) that

$$(f, A^*F(\Delta)g) = \int_\Delta t\, d(F_tf, g),$$

and since $D(A)$ is dense in H, instead of an f in $D(A)$ we can take an arbitrary vector $h \in H$.

In conclusion we may mention that there is also a theory of extensions of bounded, non-symmetric operators into normal operators with emergence from the space (such extensions are possible only when a number of conditions are satisfied). This theory and related topics are dealt with in the work of B. Sz.-Nagy and his school; see Sz.-Nagy and Foiaş [122].

§112. Spectral functions of a symmetric operator and generalized resolvents

We now bring into consideration an operator function $\boldsymbol{R}(z)$ which will play a rôle, relative to a spectral function of a symmetric operator, similar to that of the resolvent, relative to the resolution of the identity for a self-adjoint operator.

Definition. Let A be a symmetric (but not a self-adjoint) operator, let B^+ be a self-adjoint extension of A in an extension H^+ of the original space H ($H \subset H^+$), let $R^+(z)$ be the resolvent of B^+, and finally let P^+ be the operator projecting H^+ on to H. Then the operator $\boldsymbol{R}(z)$ defined for any non-real z by the formula

$$\boldsymbol{R}(z) = P^+ R^+(z)$$

in the whole space H is called the *generalized resolvent*, of the operator A, *which is generated by the self-adjoint extension* B^+.

The generalized resolvents of an operator A (with equal deficiency numbers) which are generated by its self-adjoint extensions of the first kind are called *orthogonal generalized resolvents.* They are at the same time the ordinary resolvents of the self-adjoint extensions by which they are generated.

The following theorem establishes for generalized resolvents an integral representation similar to the one obtained in Chapter 6 for the ordinary resolvents of self-adjoint operators.

Theorem 1. The operator function $\boldsymbol{R}(z)$ *will be a generalized resolvent of a symmetric operator* A *(when* $\mathcal{I}(z) \neq 0$*) if and only if it has the representation*

$$(\boldsymbol{R}(z)f, g) = \int_{-\infty}^{\infty} \frac{d(F_t f, g)}{t - z} \qquad (f, g \in H), \tag{1}$$

where F_t *is some spectral function of the operator* A.

Proof. To prove the necessity of the representation (1) we have to take the spectral function F_t which corresponds to that self-adjoint extension B^+ which generates the generalized resolvent $\boldsymbol{R}(z)$.

To prove the sufficiency of the representation (1) we have to take the generalized resolvent $\boldsymbol{R}(z)$ which is generated by that self-adjoint extension B^+ to which the spectral function F_t belongs.

If the spectral function F_t and the generalized resolvent $\boldsymbol{R}(z)$ are connected by the formula (1), then they are said to belong to one another.

By virtue of the Stieltjes inversion formula (§69) there is a one-to-one correspondence between the set of spectral functions of a given symmetric operator and the set of its generalized resolvents.

We observe that the values of all generalized resolvents of the operator A are the same on every vector in the sub-space

$$M_z = \mathrm{R}(A - zI) \qquad (\mathcal{I}(z) \neq 0).$$

For, if

$$(A - zI)f = g, \tag{2}$$

then

$$\boldsymbol{R}(z)g = \boldsymbol{R}(z)(A - zI)f = P^+R^+(z)(B^+ - zI^+)f = P^+f = f. \tag{3}$$

If the generalized resolvent $\boldsymbol{R}(z)$ is generated by the extension B^+ which has the resolvent $R^+(z)$, then, for $f, g \in \mathrm{H}$, we have

$$(\boldsymbol{R}(z)f, g) = (P^+R^+(z)f, g) = (R^+(z)f, g) = (f, R^+(\bar{z})g)$$
$$= (f, P^+R^+(\bar{z})g) = (f, \boldsymbol{R}(\bar{z})g),$$

and so

$$\boldsymbol{R}(\bar{z}) = \boldsymbol{R}^*(z). \tag{4}$$

If A is a maximal symmetric operator, and if, for example,

$$\mathrm{M}_z = \mathrm{H} \qquad \text{for} \quad \mathscr{I}(z) > 0,$$

then its generalized resolvents must coincide when $\mathscr{I}(z) > 0$, because by (2) and (3), for $\mathscr{I}(z) > 0$ and for any $g \in \mathrm{H}$,

$$\boldsymbol{R}(z)g = (A - zI)^{-1}g.$$

But, by (4), if the operators $\boldsymbol{R}(z)$ coincide for $\mathscr{I}(z) > 0$, then they coincide for $\mathscr{I}(z) < 0$.

Thus, a maximal operator has a unique generalized resolvent, and therefore it has a unique spectral function.

On the other hand, the generalized resolvent of a non-maximal symmetric operator is not uniquely defined. To see this, it suffices to choose two different maximal symmetric extensions of the first kind, C' and C'', of the operator A, and then to extend them into self-adjoint operators B' and B''. Clearly, the generalized resolvents $\boldsymbol{R}'(z)$ and $\boldsymbol{R}''(z)$ of the operators A which are generated by the extensions B' and B'' will not coincide.

Since a non-maximal symmetric operator has various generalized resolvents, its spectral function is not uniquely defined.

Remembering what was said earlier (in §111, 1°) about the spectral function of a self-adjoint operator, we see that the following result has been proved.

Theorem 2. A symmetric operator has a unique spectral function if and only if it is maximal.

This unique spectral function is orthogonal if and only if the operator is self-adjoint.

It is easy to see that the set of all spectral functions (or, the set of all generalized resolvents) of a given symmetric operator is *convex*.

This means that, if F'_t and F''_t (or $\boldsymbol{R}'(z)$ and $\boldsymbol{R}''(z)$) are two spectral functions (or two generalized resolvents) of an operator A, and if $\alpha' + \alpha'' = 1$ with $\alpha' > 0$, $\alpha'' > 0$, then the operator function $\alpha' F'_t + \alpha'' F''_t$

(or $\alpha'\boldsymbol{R}'(z)+\alpha''\boldsymbol{R}'(z)$) also is a spectral function (or a generalized resolvent) of this operator.

In this connection we indicate one method of constructing generalized resolvents (or spectral functions), which are, moreover, generated by self-adjoint extensions of the second kind by means of operations within the original space H.

Let A be a symmetric operator which operates in the space H, and let A' and A'' be any two of its maximal symmetric extensions of the first kind. We assume, for the sake of definiteness, that, for $\mathcal{I}(z)>0$

$$\mathrm{R}(A'-zI)=\mathrm{R}(A''-zI)=\mathrm{H}.$$

We define the operators $\boldsymbol{R}'(z)$ and $\boldsymbol{R}''(z)$ by

$$\boldsymbol{R}'(z)=\begin{cases}(A'-zI)^{-1} & (\mathcal{I}(z)>0),\\ (\boldsymbol{R}'(\bar{z}))^* & (\mathcal{I}(z)<0),\end{cases}$$

$$\boldsymbol{R}''(z)=\begin{cases}(A''-zI)^{-1} & (\mathcal{I}(z)>0),\\ (\boldsymbol{R}''(\bar{z}))^* & (\mathcal{I}(z)<0).\end{cases}$$

The operators $\boldsymbol{R}'(z)$ and $\boldsymbol{R}''(z)$ are generalized resolvents of the maximal symmetric operators A' and A'' respectively.

By means of the generalized resolvents $\boldsymbol{R}'(z)$ and $\boldsymbol{R}''(z)$ we find the spectral functions F'_t and F''_t of the operators A' and A'', and then form the operator function

$$F_t=\alpha' F'_t+\alpha'' F''_t \qquad (\alpha'+\alpha''=1,\ \alpha'>0,\ \alpha''>0).$$

Because the set of spectral functions is convex, this function F_t is a spectral function of the operator A.

We now show how to choose the maximal extensions A' and A'' in order that the spectral function F_t shall be generated by an extension of the first kind, or, in other words, so that the spectral function shall determine the domain D(A) of the operator A by means of the inequality

$$\int_{-\infty}^{\infty} t^2\, d(F_t f, f)<\infty. \tag{5}$$

It is easy to see that a vector $f\in\mathrm{H}$ will satisfy the condition (5) if and only if

$$f\in\mathrm{D}(A')\cap\mathrm{D}(A'').$$

It remains to prove that the maximal extensions A' and A'' of the operator A can be chosen so that the intersection of the domains D(A') and D(A'') will be the domain D(A).

By the second Neumann formula (§102)

$$\mathrm{D}(A')=\mathrm{D}(A)+\Gamma' \qquad \text{and} \qquad \mathrm{D}(A'')=\mathrm{D}(A)+\Gamma'',$$

where Γ' is the set of all vectors of the form

$$g + U'g \qquad (g \in \mathrm{N}_{\bar{z}}),$$

U' being an isometric operator mapping $\mathrm{N}_{\bar{z}}$ into $U'\mathrm{N}_{\bar{z}} \subseteq \mathrm{N}_z$ $(\mathscr{I}(z)<0$, if we assume, for definiteness, that $\mathfrak{m} \geqslant \mathfrak{n})$.

Γ'' is defined similarly.

Moreover, $\mathrm{D}(A') \cap \mathrm{D}(A'') \supseteq \mathrm{D}(A)$. We shall assume that $\mathrm{D}(A)$ is a proper sub-set of the intersection $\mathrm{D}(A') \cap \mathrm{D}(A'')$, and we let $h \notin \mathrm{D}(A)$ be a vector which belongs to both $\mathrm{D}(A')$ and $\mathrm{D}(A'')$. Then

$$h = f' + g' + U'g' \qquad \text{and} \qquad h = f'' + g'' + U''g'',$$

where $f', f'' \in \mathrm{D}(A)$ and $g', g'' \in \mathrm{N}_{\bar{z}}$. It follows from these representations that

$$(f'' - f') + (g'' - g') + (U''g'' - U'g') = 0.$$

But the bracketed terms on the left-hand side belong respectively to $\mathrm{D}(A)$, $\mathrm{N}_{\bar{z}}$, and N_z, and since these manifolds are linearly independent, it follows that

$$f' = f'', \qquad g' = g'' = g, \qquad U'g = U''g. \tag{6}$$

The last of these relations implies that, in order to satisfy the condition

$$\mathrm{D}(A') \cap \mathrm{D}(A'') = \mathrm{D}(A),$$

we must choose the isometric operators U' and U'', which determine the extensions A' and A'', in such a way that the third equality in (6) will not hold for any non-zero vector in $\mathrm{N}_{\bar{z}}$. This can always be done, for example, by choosing U' arbitrarily, and then taking $U'' = -U'$.

Thus the constructed operator functions

$$F_t = \alpha' F_t' + \alpha'' F_t'' \qquad \text{and} \qquad \boldsymbol{R}(z) = \alpha' \boldsymbol{R}'(z) + \alpha'' \boldsymbol{R}''(z)$$

with arbitrary α', α'' $(\alpha' + \alpha'' = 1, \alpha' > 0, \alpha'' > 0)$ are respectively spectral functions and generalized resolvents of the operator A which are generated, in the sense of §111, by certain self-adjoint extensions of the second kind, B^+, of the operator A.

It cannot be claimed, however, that all the spectral functions of a symmetric operator can be obtained by a method of this sort; see, in this connection Naimark [89], and Glazman and Naiman [38].

It follows, incidentally, from the method indicated for constructing spectral functions that Theorem 2 remains true even when the spectral functions are restricted to be those generated by self-adjoint extensions of the second kind.

To see this, we have to show that a non-maximal symmetric operator A has various spectral functions which are generated by self-adjoint extensions of the second kind of the operator A. But this is certainly true, because, if the maximal extensions A' and A'' are chosen as indicated above, then the generalized resolvents

$$\alpha \boldsymbol{R}'(z)+\beta \boldsymbol{R}''(z) \qquad \text{and} \qquad \beta \boldsymbol{R}'(z)+\alpha \boldsymbol{R}''(z) \qquad (\alpha \neq \beta)$$

will coincide only on vectors in $\mathrm{R}(A-zI)$, and therefore they will generate different spectral functions.

Further developments in the theory of generalized resolvents have been obtained by A. V. Shtraus; see [114, 117]. In particular, he derived a formula describing all possible generalized resolvents in terms of the space H.

In conclusion we illustrate the facts and methods discussed in this and the preceding sections by the example of the operator of differentiation.

Let $P_0 = i(d/dt)$ be the operator of differentiation, operating in the space $\mathrm{L}^2(0, \infty)$. The deficiency index of P_0 is $(0, 1)$.

To obtain a generalized self-adjoint extension of the operator P_0 by the method of Theorem 1 of §111, we construct the operator P_0' of differentiation in the space $\mathrm{L}^2(-\infty, 0)$, defining it by the formula

$$P_0' = i\frac{d}{dt} \qquad \text{on the set of functions } \varphi(t) \text{ such that:}$$

$$\{\varphi \text{ is absolutely continuous; } \varphi(t)\in \mathrm{L}^2(-\infty, 0);\ \varphi'(t)\in \mathrm{L}^2(-\infty, 0);\ \varphi(0)=0\}.$$

Clearly, the deficiency index of P_0' is $(1, 0)$.

We now form the orthogonal sums

$$\mathrm{L}^2(-\infty, \infty) = \mathrm{L}^2(-\infty, 0)\oplus \mathrm{L}^2(0, \infty),$$

$$P_0^+ = P_0'\oplus P_0 .$$

Clearly, the operator P_0^+ is defined by the formula

$$P_0^+\varphi = i\frac{d}{dt}\varphi(t) \qquad \text{on the set of functions } \varphi(t) \text{ such that:}$$

$$\{\varphi \text{ is absolutely continuous in the intervals } (-\infty, 0) \text{ and } (0, \infty);$$

$$\varphi(t)\in \mathrm{L}^2(-\infty, \infty);\ \varphi'(t)\in \mathrm{L}^2(-\infty, \infty);\ \varphi(0)=0\}.$$

It is easily seen that the domain $\mathrm{D}(P_0^+)^*$ of the adjoint operator $(P_0^+)^*$ is obtained from $\mathrm{D}(P_0^+)$ by dropping the boundary condition $\varphi(0)=0$. Hence it follows that each of the equations

$$(P_0^+)^*g + ig = 0 \qquad \text{and} \qquad (P_0^+)^*g - ig = 0$$

has a unique solution; these solutions are given, respectively, by

$$g_1(t) = \begin{cases} 0 & (t<0) \\ e^{-t} & (t\geqslant 0) \end{cases} \qquad \text{and} \qquad g_2(t) = \begin{cases} e^{t} & (t<0) \\ 0 & (t\geqslant 0). \end{cases}$$

Thus the deficiency index of the operator P_0^+ is (1, 1).

The operator P of differentiation on the real line $(-\infty, \infty)$ which was discussed in §55 is evidently a self-adjoint extension of P_0^+.

We have thus obtained, as a generalized self-adjoint extension of the operator of differentiation on the half-line, the operator of differentiation on the whole line.

§113. M. G. Krein's formula for generalized resolvents

In connection with the results of the preceding sections, there is the problem of describing the set of all the spectral functions of a given symmetric operator. Since there is a one-to-one correspondence, expressed by the formula

$$(\boldsymbol{R}(\lambda)f, g) = \int_{-\infty}^{\infty} \frac{d(F_t f, g)}{t-\lambda},$$

between the set of all spectral functions and the set of all resolvents of a symmetric operator, this problem reduces to that of describing all the resolvents. The latter problem was solved by M. G. Krein for the case where the deficiency numbers are equal and finite; see Krein [58]. Here we present M. G. Krein's results for the case where the deficiency index is (1, 1). (In a different form a formula for the generalized resolvents of an operator with the deficiency index (1, 1) was obtained by M. A. Naimark; see [88].)

Let A be a symmetric operator with the deficiency index (1, 1), let $\mathring{A}$ be some fixed self-adjoint extension of the first kind of the operator A, let $\mathring{R}(\lambda)$ be the resolvent of the operator $\mathring{A}$, and finally let $\boldsymbol{R}(\lambda)$ be an arbitrary generalized resolvent of the operator A, so that

$$\boldsymbol{R}(\lambda) = P^+ R^+(\lambda),$$

where $R^+(\lambda)$ is the orthogonal resolvent of some self-adjoint extension A^+, of the operator A, operating in an extended space $\mathrm{H}^+ \supset \mathrm{H}$, and P^+ is the operator of projection of H^+ on to H.

As usual we put

$$\mathrm{M}_\lambda = \mathrm{R}(A - \lambda I) \qquad (\mathcal{I}(\lambda) \neq 0),$$

$$\mathrm{N}_\lambda = \mathrm{H} \ominus \mathrm{M}_\lambda .$$

In the case under consideration $\quad \dim(\mathrm{N}_\lambda) = 1.$

For the difference of the resolvents we obtain, as in §106,

$$(\mathring{R}(\lambda) - \boldsymbol{R}(\lambda))f \qquad \begin{cases} = 0 & \text{for } f \in \mathrm{M}_\lambda, \\ \in \mathrm{N}_{\bar{\lambda}} & \text{for } f \in \mathrm{N}_\lambda. \end{cases} \tag{1}$$

The latter follows from the fact that, for $f \in \mathrm{N}_\lambda$ and for any $h \in \mathrm{M}_{\bar\lambda}$, by formula (4) in §112,

$$((\mathring{R}(\lambda)-\boldsymbol{R}(\lambda))f, h) = (f, (\mathring{R}(\bar\lambda)-\boldsymbol{R}(\bar\lambda))h) = (f, 0) = 0.$$

Repeating the argument of §106, we obtain from (1) the analogue of formula (6) of §106 for our present case,

$$\boldsymbol{R}(\lambda)f = \mathring{R}(\lambda)f - p(\lambda)(f, g(\bar\lambda))g(\lambda), \tag{2}$$

where

$$g(\lambda) = g(\lambda_0) + (\lambda-\lambda_0)\mathring{R}(\lambda)g(\lambda_0), \tag{2'}$$

and $g(\lambda_0)$ is a vector in $\mathrm{N}_{\bar\lambda_0}$ such that $\|g(\lambda_0\| = 1$. Here we also assume, for definiteness, that the fixed point λ_0 lies in the upper half-plane (we could take $\lambda_0 = i$). But, in general, the function $Q(\lambda) \equiv p^{-1}(\lambda)$ no longer satisfies the relation

$$Q(\lambda) = Q(\lambda_0) + (\lambda-\lambda_0)(g(\lambda_0), g(\bar\lambda)) \tag{3}$$

which we obtained in §106 (as the formula (12)) for the case of an orthogonal resolvent.

Let us investigate the nature of the function $p(\lambda)$. First of all we find $p(\lambda_0)$; to do this we put $\lambda = \lambda_0$ in (2), and go over from the resolvents to Cayley transforms

$$\mathring{U}(\lambda_0) = (\mathring{A}-\lambda_0 I)(\mathring{A}-\bar\lambda_0 I)^{-1}, \qquad U^+(\lambda_0) = (A^+ - \lambda_0 I)(A^+ - \bar\lambda_0 I)^{-1}.$$

An elementary calculation shows that

$$P^+U^+(\bar\lambda_0)f = \mathring{U}(\bar\lambda_0)f - (\lambda_0-\bar\lambda_0)p(\lambda_0)(f, g(\bar\lambda_0))g(\lambda_0),$$

and on putting $f = g(\bar\lambda_0)$ we obtain

$$P^+U^+(\bar\lambda_0)g(\bar\lambda_0) = g(\lambda_0) - (\lambda_0-\bar\lambda_0)p(\lambda_0)g(\lambda_0). \tag{4}$$

On the other hand, since the element $\psi = U^+(\bar\lambda_0)\varphi$ traverses $\mathrm{M}_{\bar\lambda_0}$ when φ traverses M_{λ_0}, we have

$$0 = (g(\bar\lambda_0), \varphi) = (U^+(\bar\lambda_0)g(\bar\lambda_0), U^+(\bar\lambda_0)\varphi) = (U^+(\bar\lambda_0)g(\bar\lambda_0), \psi)$$
$$= (P^+U^+(\bar\lambda_0)g(\bar\lambda_0), \psi),$$

which shows that

$$P^+U^+(\bar\lambda_0)g(\bar\lambda_0) = \theta g(\lambda_0),$$

where $\theta = \theta(\lambda_0)$ is a parameter, such that $|\theta| \leqslant 1$, which is determined by the extension A^+, and which satisfies $|\theta| = 1$ if and only if A^+ is an extension of the first kind.

From these considerations we obtain the important consequence that:

if the relation

$$\boldsymbol{R}(\lambda)=\mathring{R}(\lambda)$$

holds for a single non-real $\lambda=\lambda_1$, *then it holds for all non-real* λ; *the same is true for the relation*

$$\boldsymbol{R}(\lambda)g(\bar{\lambda})=\mathring{R}(\lambda)g(\bar{\lambda}).$$

For, from formula (2) we find that, for $\lambda=\lambda_1$, $p(\lambda_1)=0$, and then, by putting $\lambda_0=\lambda_1$ in (4) and (5) we obtain that the factor $\theta=\theta(\lambda_1)$ in (5) is equal to 1. So $\boldsymbol{R}(\lambda)$ is an orthogonal resolvent. But two orthogonal resolvents for the same operator are obviously identical if they coincide at one point.

Comparing (4) and (5) we find that

$$p(\lambda_0)=\frac{1-\theta}{\lambda-\bar{\lambda}_0} \quad \text{or} \quad Q(\lambda_0)\equiv p^{-1}(\lambda_0)=i\mathscr{I}(\lambda_0)+\tau, \tag{6}$$

where τ is a new parameter, connected with the parameter θ by the relation

$$\tau=i\frac{1+\theta}{1-\theta}\mathscr{I}(\lambda_0),$$

which maps the unit disc in the θ-plane on to the upper half of the τ-plane.

If $\boldsymbol{R}(\lambda)$ traverses the set of orthogonal resolvents, then, since (3) holds in this case, the formula (2) takes the form

$$\boldsymbol{R}(\lambda)f=\mathring{R}(\lambda)f-\frac{(f,g(\bar{\lambda}))g(\lambda)}{\tau+Q_1(\lambda)}, \tag{7}$$

where

$$Q_1(\lambda)=i\mathscr{I}(\lambda_0)+(\lambda-\lambda_0)(g(\lambda_0),g(\bar{\lambda})), \tag{7'}$$

and the parameter τ traverses the real line $-\infty<\tau<\infty$.

The correspondence defined by (7) between the set of all orthogonal resolvents and values of the parameter τ is one-to-one.

In the general case when $\boldsymbol{R}(\lambda)$ is an arbitrary resolvent, we put

$$p(\lambda)=Q^{-1}(\lambda)=[\tau(\lambda)+Q_1(\lambda)]^{-1}$$

in (2), and thus obtain

$$\boldsymbol{R}(\lambda)f=\mathring{R}(\lambda)f-\frac{(f,g(\bar{\lambda}))g(\lambda)}{\tau+Q_1(\lambda)}. \tag{8}$$

We shall prove that the function $\tau(\lambda)$ is holomorphic and has a non-negative imaginary part in the upper half-plane. To do this, we note that, if $\mathscr{I}(\lambda)\neq 0$,

$$\boldsymbol{R}(\lambda)g(\bar{\lambda})\neq \mathring{R}(z)g(\bar{\lambda}),$$

and also

$$(\boldsymbol{R}(\lambda)g(\bar{\lambda}),\ g(\lambda))\neq(\mathring{R}(\lambda)g(\bar{\lambda}),\ g(\lambda))$$

by the result proved above. So we put $f=g(\bar{\lambda})$ in (8) and multiply scalarly by $g(\lambda)$ to obtain

$$[\tau(\lambda)+Q_1(\lambda)]^{-1}=((\mathring{R}(\lambda)-\boldsymbol{R}(\lambda))g(\bar{\lambda}),\ g(\lambda)).$$

In this relation $Q_1(\lambda)$ and the right-hand side are holomorphic, and the latter is non-zero; therefore the function $\tau(\lambda)$ is holomorphic for every non-real λ.

Next we shall show that $\mathscr{I}(\tau(\lambda'))$ cannot be zero at even a single point of the upper half-plane unless $\tau(\lambda)=$ a constant. This will imply, *inter alia*, that the inequality $\mathscr{I}(\tau(\lambda))<0$ cannot hold at any point in the upper half-plane.

Suppose, then, that $\mathscr{I}(\tau(\lambda'))=0 \qquad (\mathscr{I}(\lambda')>0)$. We take $\lambda=\lambda'$ in both (7) and (8), and take the constant τ in (7) to be the number $\tau(\lambda')$. Then the right-hand sides of these formulae will be the same for any f. Therefore, at the point $\lambda=\lambda'$, the generalized resolvent $\boldsymbol{R}(\lambda)$ coincides with some orthogonal resolvent $R(\lambda)$, and hence, as we showed earlier, these resolvents must be identical.

By multiplying both sides of (8) scalarly by f, and bringing the right-hand side to a common denominator, we obtain the formula

$$(\boldsymbol{R}(\lambda)f,f)=\frac{p_0(\lambda)+p_1(\lambda)\tau(\lambda)}{Q_1(\lambda)+\tau(\lambda)}.$$

For a fixed f the functions $p_0(\lambda)$, $p_1(\lambda)$, and $Q(\lambda)$ do not depend on the choice of the resolvent $\boldsymbol{R}(\lambda)$ which determines the function $\tau(\lambda)$. Since the orthogonal resolvents are obtained when $\tau(\lambda)$ is a real constant, and the others when $\tau(\lambda)$ traverses some set of holomorphic functions with non-negative imaginary parts, it follows that the point $w=(\boldsymbol{R}(\lambda)f,f)$ belongs to some circular region. The boundary of this region, a circle $C(f;\lambda)$, is traced out by the point w when $\boldsymbol{R}(\lambda)$ runs through the orthogonal resolvents. In other words, the circle $C(f;\lambda)$ is described by the point

$$w=\frac{p_0(\lambda)+p_1(\lambda)\tau}{Q_1(\lambda)+\tau}$$

when τ traverses the real line.

Since the set of all resolvents of a given operator is convex, the point $w = (\boldsymbol{R}(\lambda)f, f)$ runs over the disc $K(f; \lambda)$ with boundary $C(f; \lambda)$ when $\boldsymbol{R}(\lambda)$ runs through the set of all resolvents of the operator A.

The preceding argument has shown that:

every resolvent $\boldsymbol{R}(\lambda)$ *of a symmetric operator* A *with deficiency index* (1, 1) *can be expressed in the form*

$$\boldsymbol{R}(\lambda) = \mathring{R}(\lambda) - \frac{(\cdot, g(\bar{\lambda}))g(\lambda)}{\tau(\lambda) + Q_1(\lambda)}, \tag{9}$$

where $\tau(\lambda)$ *belongs to the class* N *of functions which are defined and holomorphic in the half-plane* $\mathscr{I}(\lambda) > 0$ *and which have a non-negative imaginary part in that half-plane.*

We now prove the converse, *viz.*, *every function* $\tau(\lambda)$ *in the class* N *generates, by the formula* (9), *a resolvent of the operator* A.

Let $\tau(\lambda) \in N$, and let $S(\lambda)$ be the operator defined by

$$S(\lambda) = \mathring{R}(\lambda) - \frac{(\cdot, g(\bar{\lambda}))g(\lambda)}{\tau(\lambda) + Q_1(\lambda)}. \tag{9'}$$

Since the circle $C(f; \lambda)$ lies in the upper half-plane and the point $(S(\lambda)f, f)$ lies in the disc $K(f; \lambda)$, the scalar product $(S(\lambda)f, f)$ — which in the half-plane $\mathscr{I}(\lambda) > 0$ is a holomorphic function — clearly belongs to the class N.

Also, because, for orthogonal resolvents $R(\lambda)$, the scalar product $(R(\lambda)f, f)$ satisfies the inequality

$$|(R(\lambda)f, f)| \leqslant \frac{(f, f)}{\mathscr{I}(\lambda)} \qquad (\mathscr{I}(\lambda) > 0),$$

the inequality

$$|(S(\lambda)f, f)| \leqslant \frac{(f, f)}{\mathscr{I}(\lambda)}$$

will also hold for the scalar product $(S(\lambda)f, f)$, lying in the disc $K(f; \lambda)$.

Using this inequality and repeating the argument of §74, we arrive at a representation for the scalar product $(S(\lambda)f, f)$ in the form

$$(S(\lambda)f, f) = \int_{-\infty}^{\infty} \frac{d\omega(t; f)}{t - \lambda}, \tag{10}$$

where $\omega(t; f) = \omega(t)$ is a non-decreasing, left-continuous function which tends to 0 as $t \to -\infty$ and which satisfies the condition

$$\omega(t; f) \leqslant (f, f) \qquad (-\infty < t \leqslant \infty).$$

From (10), by repeating the argument of §74, we obtain a representation for the operator $S(\lambda)$ in the form

$$(S(\lambda)f, g) = \int_{-\infty}^{\infty} \frac{d(F_t f, g)}{t-\lambda}, \qquad (10')$$

where F_t is a non-decreasing, left-continuous operator function, which tends to 0 as $t \to -\infty$ and which satisfies the condition

$$(F_t f, f) \leq (f, f) \qquad (-\infty < t \leq \infty). \qquad (11)$$

But now, in contrast to §74, the operator function $S(\lambda)$ does not satisfy the Hilbert functional relation, and so we cannot derive for F_t the orthogonality property

$$F_u F_v = F_s \qquad (s = \min\{u, v\}). \qquad (12)$$

We cannot now use (12) to prove the property

$$\lim_{t\to\infty} F_t f = f \qquad (f \in \mathrm{H}), \qquad (13)$$

which we obtained at the end of §74.

To prove that the operator function F_t is a generalized resolution of the identity we have to prove that (13), is true, without using the relation (12).

The limit in (13) certainly exists—by the theorem on monotone increasing sequences of operators (§33). It therefore suffices to show that, as $t \to \infty$, the operator F_t tends weakly to I, that is, that for any f and g in H,

$$\lim_{t\to\infty} (F_t f, g) - (f, g).$$

Clearly, this relation will hold if, for every $f \in \mathrm{H}$,

$$\lim_{t\to\infty} (F_t f, f) = (f, f). \qquad (14)$$

But, by (11), the norm of the operator function F_t is bounded ($\|F_t\| \leq 1$), and therefore it suffices to verify that (14) holds for some set of vectors which is dense in H. We take as this set the domain $\mathrm{D}(A)$ of the operator A

By (10′) the relation to be proved, (14), is equivalent to the relation

$$\lim_{\eta\to\infty} i\eta(S(i\eta)f, f) = -(f, f). \qquad (15)$$

Since the point $(S(i\eta)f, f)$ lies in the disc $K(f; i\eta)$, it is clear that

$$|i\eta(S(i\eta)f, f) + (f, f)| \leq \max_{-\infty < \tau \leq \infty} |i\eta(R(i\eta; \tau)f, f) + (f, f)|,$$

where $R(\lambda;\tau)$ denotes the resolvent of the first kind corresponding to the parameter τ. It remains for us to prove that, as $\eta\to\infty$, the quantity

$$|i\eta(R(i\eta;\tau)f,f)+(f,f)|$$

tends to 0 uniformly with respect to τ $(-\infty<\tau\leqslant\infty)$. But this is true, because, if $f\in D(A)$ and if $\eta>0$,

$$|i\eta(R(i\eta;\tau)f,f)+(f,f)|$$

$$=\left|\int_{-\infty}^{\infty}\frac{i\eta}{t-i\eta}\,d(E_t^\tau f,f)+\int_{-\infty}^{\infty}d(E_t^\tau f,f)\right|=\left|\int_{-\infty}^{\infty}\frac{t}{t-i\eta}\,d(E_t^\tau f,f)\right|$$

$$\leqslant\sqrt{\left\{\int_{-\infty}^{\infty}t^2d(E_t^\tau f,f)\right\}}\cdot\sqrt{\left\{\int_{-\infty}^{\infty}\frac{1}{|t-i\eta|^2}\,d(E_t^\tau f,f)\right\}}\leqslant\frac{1}{\eta}\|Af\|\,.\,\|f\|.$$

This proves the relation (13), and thus the operator function F_t is indeed a generalized resolution of the identity, and therefore, by Naimark's theorem, it has the representation

$$F_t=P^+E_t^+.$$

We now introduce the self-adjoint operator A^+ defined by

$$A^+f=\int_{-\infty}^{\infty}t\,dE_t^+f.$$

Clearly,

$$S(\lambda)=P^+R^+(\lambda),$$

where $R^+(\lambda)$ is the resolvent of A^+, and to complete the proof of the theorem it remains to show that A^+ is an extension of the operator A, *i.e.*, that

$$A^+\supset A,$$

or that

$$R^+(\lambda)f=\boldsymbol{R}(\lambda)f \qquad \text{for} \quad f\in M_\lambda.$$

It follows from (9′) that, for $f\in M_\lambda$,

$$P^+R^+(\lambda)f=S(\lambda)f=\mathring{R}(\lambda)f,$$

and so

$$R^+(\lambda)f=\boldsymbol{R}(\lambda)f+h,$$

where $h\perp H$. We shall prove that $h=0$. If $f\in M_\lambda$, we have

$$A\boldsymbol{R}(\lambda)f=f+\lambda\boldsymbol{R}(\lambda)f,$$

$$A^+(\boldsymbol{R}(\lambda)f+h)=f+\lambda\boldsymbol{R}(\lambda)f+\lambda h,$$

from which, on putting $g = \boldsymbol{R}(\lambda)f$, we obtain

$$(A^+(g+h), g+h) = (Ag + \lambda h, g+h) = (Ag, g) + \lambda(h, h),$$

and since $\mathscr{I}(\lambda) \neq 0$, this is possible only if $h = 0$.

The theorem has been proved completely.

§114. Quasi-self-adjoint extensions and the characteristic function of a symmetric operator

In this section we discuss another form, introduced by M. S. Livshits, of extensions of symmetric operators with finite and equal deficiency numbers; see [75, 76].

A *quasi-self-adjoint extension* of a symmetric operator A with deficiency index (m, m) is a linear operator B which satisfies the conditions

$$A \subset B \subset A^*, \tag{1}$$

$$\dim D(B) = m \qquad (\operatorname{mod} D(A)), \tag{2}$$

but which is not a self-adjoint extension of A.

To simplify the presentation, we restrict ourselves to the case of operators with deficiency index (1, 1), in which case the condition (2) is a consequence of condition (1), and can therefore be dropped.

We shall assume that the operator A is simple (§103).

Example of a quasi-self-adjoint extension of a symmetric operator.

Let P be the operator of differentiation in the space $L^2(0, a)$ with the boundary conditions

$$\varphi(0) = 0 = \varphi(a).$$

In Chapter 4 we showed that the domain of any self-adjoint extension P_θ of this operator P is defined by the boundary conditions

$$\varphi(a) = \theta\varphi(0) \qquad (|\theta| = 1), \tag{3}$$

and, conversely, every boundary condition of this kind with $|\theta| = 1$ determines the domain of some self-adjoint extension of P.

If in equation (3) we now take, instead of θ, an arbitrary complex number ρ with $|\rho| \neq 1$, then the operator P_ρ $(P_\rho \subset P^*)$, which is defined on the set of functions $\varphi(t)$ such that

$$\varphi(a) = \rho\varphi(0) \qquad (|\rho| \neq 1) \tag{4}$$

will be a quasi-self-adjoint extension of the operator P. Also it is clear that every quasi-self-adjoint extension of the operator P is specified by condition (4) with some ρ (one of them is given by $p=\infty$).

For quasi-self-adjoint extensions B of an operator A we can introduce the Cayley transform S, defined on the manifold $D(S)=R(B-\bar{\lambda}I)$ by the formulae

$$\varphi=(B-\bar{\lambda}I)f, \qquad (f\in D(B), \mathcal{I}(\lambda)\neq 0). \tag{5}$$

$$S\varphi=(B-\lambda I)f, \tag{6}$$

This definition makes sense unless the vector f is not determined uniquely by φ, *i.e.*, unless $\bar{\lambda}$ is an eigenvalue of the operator B. But if $\bar{\lambda}$ is an eigenvalue of B, then λ is not an eigenvalue of B (for if it were, we would have

$$g_\lambda\in D(B), \qquad g_{\bar{\lambda}}\in D(B),$$

and also $B=A^*$, contradicting the condition (1)).

We may therefore assume that, in (5), $\bar{\lambda}$ is not an eigenvalue of the operator B. For definiteness we take $\lambda=i$, and then (5) and (6) become

$$\varphi=(B+iI)f, \qquad (f\in D(B)). \tag{5'}$$

$$S\varphi=(B-iI)f, \tag{6'}$$

It is clear that the operator S is defined on the whole space, and that it is an extension of the Cayley transform V of the operator A. But now, in contrast to the case of self-adjoint extensions, the operator S will not be unitary.

We shall prove that the orthogonal complement of the manifold

$$D(V)=M_{-i}=R(A-iI)$$

is mapped by S into the orthogonal complement of the manifold

$$R(V)=M_i=R(A+iI).$$

For, if $g\perp M_{-i}$, then, putting in accordance with (5')

$$g=(B+iI)h \qquad (h\in D(B))$$

we have

$$((B+iI)h,(A+iI)f)=0 \qquad \text{for any } f\in D(A).$$

On the other hand, for any $f\in D(A)$, we find by using the inclusion $B\subset A^*$ that

$$\begin{aligned}(Sg,(A-iI)f)&=((B-iI)h,(A-iI)f)\\&=(Bh,Af)-i(h,Af)+i(Bh,f)+(h,f)\\&=(Bh,Af)-i(Bh,f)+i(h,Af)+(h,f)=((B+iI)h,(A+iI)f)=0,\end{aligned}$$

and so $Sg\perp M_i$ (in particular, Sg may be equal to 0).

If g^* is a vector orthogonal to the manifold M_i and such that $\|g^*\| = \|g\|$, then, from what has been proved,

$$Sg = \kappa g^* \tag{7}$$

where the number κ is not equal to 1 in absolute value, for if it were, then the operator B, which, clearly, can be expressed in terms of S by means of the formulae

$$f = \frac{1}{2i}(I - S)\varphi, \tag{8}$$

$$(\varphi \in H),$$

$$Bf = \frac{1}{2}(I + S)\varphi, \tag{9}$$

would be self-adjoint.

The operator $S \equiv S_\kappa$ is said to be a *quasi-unitary extension* of the isometric operator V. It is easily verified that, if $\kappa \neq 0$, then

$$S_\kappa^* = S_\eta^{-1}, \qquad \text{where} \qquad \eta = \bar{\kappa}^{-1}.$$

In general, a *quasi-unitary extension* of a given isometric operator V with the deficiency index (m, m) $(m < \infty)$ is any linear (but not unitary) operator $S \supset V$ which is defined in the whole space and which maps the orthogonal complement of the manifold $D(V)$ into some sub-space $F \subseteq H \ominus R(V)$.

Clearly, every quasi-unitary extension S of the operator V generates, through the formulae (8) and (9), a quasi-self-adjoint extension B of the operator A.

When the deficiency index is (1, 1), there is a one-to-one correspondence between the set of complex numbers κ with $|\kappa| \neq 1$ and the set of all quasi-self-adjoint extensions B of the operator A (or the set of quasi-unitary extensions S of the operator V). We shall therefore denote such extensions by B_κ (or S_κ).

We now investigate the spectrum of a quasi-self-adjoint extension B of an operator A with deficiency index (1, 1).

We recall that, in accordance with the general definition given in §48, a number λ is called a regular point of a linear operator T if the operator $(T - \lambda I)^{-1}$ exists, is bounded, and is defined in the whole space. The complement of the set of regular points of the operator T is called its spectrum.

Theorem 1. *The spectrum of any quasi-self-adjoint extension B of a simple operator A with deficiency index (1, 1) consists of the spectral kernel (see §105) of the operator A and the eigenvalues of A. The entire set of eigenvalues lies either in the upper half-plane, or in the lower half-plane. If* *the one exceptional case (see the corollary to Theorem* 3) *when the whole (upper or lower) half-plane consists of eigenvalues is excluded, then the set of eigenvalues can have only real limit-points.*

Proof. Let Λ be the spectral kernel of the operator A. For any $\lambda \in \Lambda$ the operator $(A-\lambda I)^{-1}$ is unbounded, and therefore the operator $(B-\lambda I)^{-1}$ cannot be bounded. Hence Λ is included in the spectrum of the operator B.

Now suppose that $\lambda \notin \Lambda$. If the operator $(B-\lambda I)^{-1}$ does not exist, then λ is an eigenvalue of B. But if the operator $(B-\lambda I)^{-1}$ does exist, then λ is a regular point of the operator B. For, the operator $(B-\lambda I)^{-1}$ cannot be unbounded, since the operator $(A-\lambda I)^{-1}$ is bounded, and $\mathrm{R}(B-\lambda I)$ differs from $\mathrm{R}(A-\lambda I)$ by at most one dimension. It remains to prove that $\mathrm{R}(B-\lambda I)=\mathrm{H}$. If we assume the contrary, then we find that $\mathrm{R}(B-\lambda I) = \mathrm{R}(A-\lambda I)$. If we now choose a vector $f \in \mathrm{D}(B)$ which does not belong to $\mathrm{D}(A)$, then the vector

$$(B-\lambda I)f=f^*,$$

which belongs to $\mathrm{R}(B-\lambda I)=\mathrm{R}(A-\lambda I)$ can be expressed as

$$f^*=(A-\lambda I)f'=(B-\lambda I)f' \qquad (f' \in \mathrm{D}(A)).$$

Therefore

$$(B-\lambda I)(f-f')=0.$$

contradicting the existence of the operator $(B-\lambda I)^{-1}$; and so the regularity of the point λ has been proved.

Thus, the spectrum of the operator B consists of the spectral kernel Λ of the operator A and the eigenvalues of A.

By bringing in the Cayley transform S_κ of the operator $B=B_\kappa$, we easily find that, as the point λ traverses the spectrum of the operator B, the point

$$\zeta=\frac{\lambda-i}{\lambda+i} \tag{10}$$

traverses the spectrum of the operator S_κ, and conversely; further, the formula (10) establishes a one-to-one correspondence between the eigenvalues of B and S_κ.

From what we have said, it suffices to find out how the eigenvalues of S_κ are disposed.

Since the operator A is simple, its Cayley transform V has no eigenvalues, and therefore, as is easily verified, if $|\kappa|<1$ the eigenvalues of S_κ are <1 in absolute value, and if $|\kappa|<1$ they are greater than 1 in absolute value. Therefore, all the eigenvalues of the operator B lie in the upper half-plane if $|\kappa|<1$, or in the lower half-plane if $|\kappa|>1$.

We shall assume that $|\kappa|<1$. In the following, let g be a vector orthogonal to the manifold M_{-i} with $\|g\|=1$, let $\mathring{U}$ be a unitary extension of the operator V, let g^* be the vector defined by

$$g^* = \mathring{U}g,$$

and let S_κ be the quasi-unitary extension of the operator V determined by the equality

$$S_\kappa g = \kappa g^*.$$

We express the eigenvector corresponding to the eigenvalue ζ of the operator S_κ in the form $\varphi + \alpha g$ $(\varphi \in D(V))$. Thus

$$S_\kappa(\varphi + \alpha g) = \zeta(\varphi + \alpha g),$$

which gives

$$S_\kappa \varphi + \alpha\kappa g^* = \zeta\varphi + \alpha\zeta g,$$

or

$$\mathring{U}\varphi - \zeta\varphi = \alpha(\zeta g - \kappa g^*),$$

which implies

$$\frac{1}{\alpha}\varphi = \zeta(\mathring{U} - \zeta I)^{-1} g - \kappa(\mathring{U} - \zeta I)^{-1} g^*. \tag{11}$$

Multiplying scalarly by g throughout (11), we obtain an equation which is satisfied by the eigenvalues of the operator S_κ, *viz.*,

$$\zeta((\mathring{U} - \zeta I)^{-1} g, g) - \kappa((\mathring{U} - \zeta I)^{-1} g^*, g) = 0. \tag{12}$$

Reversing the steps of the argument shows that every root of this equation is an eigenvalue of the operator S_κ.

Since the left-hand side of equation (12) is a regular function when $|\zeta| \neq 1$, the assertion of the theorem about the limit-points of the discrete spectrum of the operator B is proved, except for the case where the left-hand side of (12) vanishes identically when $|\zeta|<1$.

In the latter case, every point ζ of the interior of the unit disc is an eigenvalue of the operator S_κ.

We shall show next that this exceptional case, which was mentioned in the statement of the theorem, can actually occur.

Let H be a separable Hilbert space with an orthonormal basis $\{e_k\}_{-\infty}^{\infty}$, and let $\tilde{U}$ be the unitary operator defined by the formulae

$$\tilde{U}e_k = e_{k-1}.$$

Suppose also that $\tilde{V} \subset \tilde{U}$ and $D(\tilde{V}) = H \ominus \{e_0\}$, and therefore

$$R(\tilde{V}) = H \ominus \{e_{-1}\}.$$

$\tilde{V}$ is an isometric operator with deficiency index (1, 1).

Now let S_0 be the quasi-unitary extension, of the operator V, defined by the condition

$$S_0 e_0 = 0.$$

It is easily verified that, for any ζ in the disc $|\zeta|<1$, the vector

$$\sum_{k=0}^{\infty} \zeta^k e_k$$

is the eigenvector corresponding to the eigenvalue ζ of the operator S_0.

We shall see later that this exceptional case is unique, in the sense that if a simple symmetric operator with deficiency index (1, 1) has a quasi-self-adjoint extension with a point spectrum which occupies a whole half-plane, then this operator is isomorphic to the Cayley transform of the operator $\tilde{V}$.

We now study a transformation of the equation (12).

If F_t is the resolution of the identity for the operator $\mathring{U}$, then

$$2\zeta((\mathring{U}-\zeta I)^{-1}g, g) = 2\zeta\int_0^{2\pi} \frac{1}{e^{is}-\zeta}\, d(\mathring{F}_s g, g) = \int_0^{2\pi} \frac{e^{is}+\zeta}{e^{is}-\zeta}\, d(\mathring{F}_s g, g) - 1, \quad (13)$$

and

$$2((\mathring{U}-\zeta I)^{-1}g^*, g) = 2((\mathring{U}-\zeta I)^{-1}U_0 g, g) = 2\int_0^{2\pi} \frac{e^{is}}{e^{is}-\zeta}\, d(\mathring{F}_s g, g) = \int_0^{2\pi} \frac{e^{is}+1}{e^{is}-1}\, d(\mathring{F}_s g, g) + 1. \quad (14)$$

(14) implies, in particular, that the scalar product $((\mathring{U}-\zeta I)^{-1}g^*, g)$ does not vanish when $|\zeta|<1$, and therefore the equation (12) can be put into the form

$$w(\zeta) - \kappa = 0,$$

where

$$w(\zeta) = \frac{\zeta((\mathring{U}-\zeta I)^{-1}g, g)}{((\mathring{U}-\zeta I)^{-1}g^*, g)} \qquad (g^* = \mathring{U}g). \quad (15)$$

By (13) and (14) the function $w(\zeta)$ can be expressed as

$$w(\zeta) = \frac{\Phi(\zeta)-1}{\Phi(\zeta)+1}, \quad (15a)$$

where

$$\Phi(\zeta) = \int_0^{2\pi} \frac{e^{is}+\zeta}{e^{is}-\zeta}\, d(\mathring{F}_s g, g). \quad (15b)$$

From these representations it follows (see §69) that the function $w(\zeta)$ is regular in the unit disc, it maps the latter into a part of that disc, and it satisfies the normalization condition $w(0)=0$.

Following M. S. Livshits, we shall call $w(\zeta)$ the *characteristic function of the isometric operator* V, and the function

$$\omega(\lambda)=w\left(\frac{\lambda-i}{\lambda+i}\right) \tag{15c}$$

the *characteristic function of the symmetric operator* A.

The function $\omega(\lambda)$ is regular in the upper half-plane, maps it into part of the unit disc, and satisfies the normalization condition $\omega(i)=0$.

To justify this definition we ought to show that although the function $w(\zeta)$ seems from (15) to depend on the choice of the unitary extension $\mathring{U}$, it really depends only on the operator V. To do this, we express $\mathring{U}$ in (15) by its Cayley transform

$$\mathring{U}=(\mathring{A}-iI)(\mathring{A}+iI)^{-1}.$$

After an elementary calculation we obtain for the characteristic function $\omega(\lambda)$ the formula

$$\omega(\lambda)=\frac{\lambda-i}{\lambda+i}\cdot\frac{((I+(\lambda-i)\mathring{R}(\lambda))g,\, g^*)}{((I+(\lambda-i)\mathring{R}(\lambda))g,\, g)}, \tag{16}$$

where $\mathring{R}(\lambda)$ is the resolvent of the operator $\mathring{A}$.

It was shown in §106 (see formula (10)) that the vector

$$(I+(\lambda-i)\mathring{R}(\lambda))g$$

belongs to the deficiency sub-space $\mathrm{N}_{\bar{\lambda}}$, and since this is 1-dimensional in the present case, this vector is merely multiplied by a scalar if some other extension $\mathring{A}$ is chosen; this would have no effect on $\omega(\lambda)$. As regards g^*, it would be multiplied by some number equal to 1 in absolute value if a different extension $\mathring{A}$ were chosen.

Thus the characteristic function of a symmetric operator (or of an isometric operator) is determined by this operator to within an arbitrary constant factor equal to 1 in absolute value. We shall not regard two characteristic functions which differ only by such a factor as being distinct.

Formula (16) can now be rewritten as

$$\omega(\lambda)=\frac{\lambda-i}{\lambda+i}\cdot\frac{(g_\lambda,\, g^*)}{(g_\lambda,\, g)}, \tag{17}$$

where g_λ is an arbitrary solution of the equation

$$A^*h-\lambda h=0.$$

As an illustration we calculate the characteristic function of the operator P of differentiation on an interval $[0, a]$ with the boundary conditions

$$\varphi(0)=\varphi(a)=0.$$

In this case

$$g=\frac{\sqrt{2}}{\sqrt{(e^{2a}-1)}}e^{t}, \qquad g^{*}=\frac{\sqrt{2}}{\sqrt{(1-e^{-2a})}}e^{-t}, \qquad g_{\lambda}=e^{-i\lambda t},$$

and by formula (17)

$$\omega(\lambda)=\frac{e^{a}-e^{-ia\lambda}}{1-e^{a}e^{-ia\lambda}}. \tag{18}$$

In addition to $w(\zeta)$ and $\omega(\lambda)$ we also introduce the functions

$$w(\zeta;\kappa)=\frac{w(\zeta)-\kappa}{\bar{\kappa}w(\zeta)-1}, \qquad \omega(\lambda;\kappa)=\frac{\omega(\lambda)-\kappa}{\bar{\kappa}\omega(\lambda)-1}$$

and call them the *characteristic functions* of a quasi-unitary extension S_{κ} of the operator V and of a quasi-self-adjoint extension B_{κ} of the operator A respectively.

The characteristic functions $w(\zeta;\kappa)$ and $\omega(\lambda;\kappa)$ are normalized by the conditions

$$w(0;\kappa)=\kappa \qquad \text{and} \qquad \omega(i;\kappa)=\kappa.$$

The spectral kernel of the operator A is also determined by the characteristic function $\omega(\lambda)$, as the next theorem shows.

Theorem 2. A real number λ_0 is a point of regular type for a simple symmetric operator A with deficiency index $(1, 1)$ if and only if both the following conditions hold:

1°. *the function $\omega(\lambda)$ be regular in a neighbourhood of λ_0;*

2°. *the function $\omega(\lambda)$ be equal to 1 in absolute value in some interval $\lambda_0-\varepsilon<\lambda<\lambda_0+\varepsilon$.*

Proof. Let λ_0 be a point of regular type for the operator A, and let V be the Cayley transform of A.

It is clear that there is a self-adjoint extension $\mathring{A}\supset A$ for which λ_0 is a regular point, because if two self-adjoint extensions $\mathring{A}_1$ and $\mathring{A}_2$ have an eigenvalue λ_0 in common, then

$$\mathring{A}_1 f_1=\lambda_0 f_1, \qquad \mathring{A}_2 f_2=\lambda_0 f_2 \qquad (f_1, f_2\notin D(A)),$$

whence

$$\dot{A}^{*}f_1=\lambda_0 f_1, \qquad A^{*}f_2=\lambda_0 f_2,$$

and therefore

$$f_1 = \alpha f_2,$$

and so

$$D(\mathring{A}_1) = D(\mathring{A}_2), \qquad i.e., \qquad \mathring{A}_1 = \mathring{A}_2.$$

If now in (15) we take as $\mathring{U}$ the Cayley transform of the operator A, then it follows from (15a, b, c) that the function $\omega(\lambda)$ has the properties 1° and 2°.

Conversely, suppose that the function $\omega(\lambda)$ has the properties 1° and 2°. Then, choosing the extension $\mathring{U}$ in (15) so that

$$w(\zeta_0) \neq 1 \qquad \left(\zeta_0 = \frac{\lambda_0 - i}{\lambda_0 + i}\right),$$

we obtain the result that the function

$$\Phi(\zeta) = \frac{1 + w(\zeta)}{1 - w(\zeta)}$$

is regular in a neighbourhood of the point $\zeta_0 = e^{is_0}$, and also that $\mathcal{R}(\Phi(\zeta)) = 0$ holds on some arc, of the unit circle, which contains the point ζ_0.

It follows from this, after applying the inversion formula of §69 to the representation (15b), that s_0 is a point of constancy of the function $(\mathring{F}_s g, g)$, and since g is a generating vector for the operator $\mathring{U}$ (see §103), so ζ_0 is a regular point for this operator (see §83). Thus λ_0 is a regular point for the operator $\mathring{A}$ and therefore it does not belong to the spectral kernel of the operator A. Q.E.D.

From Theorem 2 we obtain the following more precise version of the statement in Theorem 1.

The finite limit-points of the set of eigenvalues of any quasi-self-adjoint extension B_κ of a given operator A belong to the spectral kernel of A (except in the special case mentioned in Theorem 1).

If ζ_k $(k = 1, 2, 3, \ldots)$ are the roots of the characteristic function $w(\zeta; \kappa)$ of the quasi-unitary extension S_κ of the operator V when $\kappa \neq 0$, then $w(\zeta; \kappa)$ can be expressed as a Blasche product

$$\omega(\zeta; \kappa) = \exp[-G(\zeta)] \prod_{k=1}^{\infty} \left\{\frac{\zeta_k - \zeta}{1 - \bar{\zeta}_k \zeta} \cdot \frac{|\zeta_k|}{\zeta_k}\right\},$$

where $G(\zeta)$ is a function which is regular when $|\zeta| < 1$ and which has a non-negative real part.

Putting $G(\zeta)$ into the form (see §69)

$$G(\zeta)=\int_0^{2\pi}\frac{e^{is}+\zeta}{e^{is}-\zeta}\,d\rho(s)+i\beta,$$

where $\rho(s)$ is a non-decreasing function of bounded variation, we obtain†

$$w(\zeta;\kappa)=C\exp\left[\int_0^{2\pi}\frac{\zeta+e^{is}}{\zeta-e^{is}}\,d\rho(s)\right]\prod_{k=1}^{\infty}\left\{\frac{1-\zeta/\zeta_k}{1-\zeta\bar{\zeta}_k}|\zeta_k|\right\}\qquad (|C|=1). \tag{19}$$

It follows from Theorems 1 and 2 that the point spectrum of a quasi-unitary extension S_κ of the operator V consists of the points ζ_k, and the other part of the spectrum consists of the points of growth of the function $\rho(s)$ and the limit-points of the set of roots ζ_k.

If the spectral kernel of an operator A is void, then the eigenvalues of any quasi-self-adjoint extension B_κ of the operator A can have a limit-point only at infinity.

In particular it may happen that the spectrum of some quasi-self-adjoint extension of the operator A is the empty set. An example of this sort is the quasi-self-adjoint extension of the operator of differentiation P defined by the boundary condition $\varphi(0)=0$.

We shall see later (from Theorem 4) that this exceptional case is unique if we do not regard isomorphic operators as distinct.

We now prove a general theorem which shows that the characteristic function defines the operator up to an isomorphism.

Theorem 3. *Simple, symmetric (or isometric) operators with deficiency index* (1, 1) *are unitarily equivalent if and only if their characteristic functions coincide.*

Proof. We give the proof for isometric operators.

Let the operators V and $\tilde{V}$, operating in the spaces H and $\tilde{\text{H}}$ respectively, be unitarily equivalent, *i.e.*, let

$$\tilde{V}=\mathfrak{U}V\mathfrak{U}^{-1},$$

where $\mathfrak{U}$ is an isometric operator mapping H on to $\tilde{\text{H}}$.

Having chosen some unitary extension U of the operator V, we construct a unitary extension $\tilde{U}$ of the operator $\tilde{V}$ by the formula

$$\tilde{U}=\mathfrak{U}U\mathfrak{U}^{-1}.$$

† The representation (19) was the starting-point for the construction of the so-called 'triangular models' for some classes of non-self-adjoint operators (see M. S. Livshits [77], and also the review article [by Brodsky and Livshits [20]).

Further, we choose a vector g (with $\|g\|=1$) belonging to H and orthogonal to the manifold $D(V)$, and we put

$$\tilde{g}=\mathfrak{U}g.$$

For this choice of the vector $\tilde{g}$ and the operator $\tilde{U}$, it follows from the formulae

$$w(\zeta)=\frac{\zeta((U-\zeta I)^{-1}g,\, g)_{\mathrm{H}}}{((U-\zeta I)^{-1}Ug,\, g)_{\mathrm{H}}} \tag{20}$$

$$\tilde{w}(\zeta)=\frac{\zeta((\tilde{U}-\zeta\tilde{I})^{-1}\tilde{g},\, \tilde{g})_{\tilde{\mathrm{H}}}}{((\tilde{U}-\zeta\tilde{I})^{-1}\tilde{U}\tilde{g},\, \tilde{g})_{\tilde{\mathrm{H}}}} \tag{21}$$

that

$$\tilde{w}(\zeta)=w(\zeta).$$

Conversely, suppose that the characteristic functions $\tilde{w}(\zeta)$ and $w(\zeta)$, of the operators $\tilde{V}$ and V, as defined by (20) and (21), coincide. Then, by putting

$$\tilde{\Phi}(\zeta)=\frac{1+\tilde{w}(\zeta)}{1-\tilde{w}(\zeta)} \qquad \text{and} \qquad \Phi(\zeta)=\frac{1+w(\zeta)}{1-w(\zeta)},$$

and applying the inversion formulae to the representations of the functions $\tilde{\Phi}(\zeta)$ and $\Phi(\zeta)$, we obtain

$$(\tilde{E}_t\tilde{g},\, \tilde{g})_{\tilde{\mathrm{H}}}=(E_tg,\, g)_{\mathrm{H}}.$$

But since the operators $\tilde{V}$ and V are simple, it follows that the spectra of their unitary extensions $\tilde{U}$ and U are simple, and also that the deficiency vectors $\tilde{g}$ and g are generating vectors for $\tilde{U}$ and U (*cf.* §103).

Thus the operators $\tilde{U}$ and U are reduced to the same canonical form, namely, to the operator of multiplication by e^{it} in the space L^2_σ with the distribution function

$$\sigma(t)=(\tilde{E}_t\tilde{g},\, \tilde{g})_{\tilde{\mathrm{H}}}=(E_tg,\, g)_{\mathrm{H}}$$

(§83), and they are therefore isomorphic. But if $\tilde{U}$ and U are isomorphic, then so are $\tilde{V}$ and V. Hence the theorem.†

Corollary. *Any simple, symmetric operator with deficiency index* (1, 1) *which admits a quasi-self-adjoint extension that has a point spectrum*

† For later generalizations of the concept of the characteristic function and the formulation of the condition for unitary equivalence in terms of it, see in addition to the papers mentioned on p. 422, Shtraus [116], and Brodsky [19].

covering the whole half-plane is isomorphic to the Cayley transform of the operator $\tilde{V}$, *defined on* *p..417.*

For, the characteristic functions of such operators must be identically zero, and so all these operators are isomorphic.

The next theorem gives a curious, abstract characterization of the operator of differentiation on a finite interval.

Theorem 4. *Any simple, symmetric operator with deficiency index* (1, 1) *which admits a quasi-self-adjoint extension that has no spectrum is isomorphic to the operator of differentiation on a finite interval.*

Proof. Let A be an operator having the properties stated in the theorem, let B_κ be a quasi-self-adjoint extension of A without a spectrum, and let $\omega(\lambda; \kappa)$ be the characteristic function of the operator B_κ.

Further, let V be the Cayley transform of the operator A, and let $w(\zeta; \kappa)$ be the characteristic function of the Cayley transform S_κ of the operator B_κ.

Since the operator B_κ has no eigenvalues, the function $w(\zeta; \kappa)$—which maps the unit circle into a part of itself—has no zeros, and can therefore be expressed as

$$w(\zeta; \kappa) = e^{-G(\zeta)},$$

where $G(\zeta)$ is a function which is regular in the unit disc and has a non-negative real part there.

Replacing ζ by $(\lambda - i)/(\lambda + i)$, we have

$$\omega(\lambda; \kappa) = e^{iH(\lambda)},$$

where $H(\lambda)$ is a function which is regular in the upper half-plane and maps that region into itself.

As in §69 the function H can be expressed as

$$H(\lambda) = \alpha + \mu\lambda + \int_{-\infty}^{\infty} \frac{1 + t\lambda}{t - \lambda}\, d\sigma(t),$$

where α is a real number, μ is $\geqslant 0$, and $\sigma(t)$ is a non-decreasing function of bounded variation on $(-\infty, \infty)$.

Further, since the spectral kernel of the operator A is void, the function $H(\lambda)$ is regular and real-valued on the real axis. Therefore, by the Stieltjes inversion formula (§60)

$$\sigma(t) = \text{constant} \qquad \text{and} \qquad H(\lambda) = \alpha + \mu\lambda.$$

Thus, if we omit the constant factor equal to 1 in absolute value, we have

$$\omega(\lambda;\kappa)=e^{i\mu\lambda},$$

whence

$$\omega(\lambda)=\frac{\kappa-\omega(\lambda;\kappa)}{1-\bar{\kappa}\omega(\lambda;\kappa)}=\frac{\kappa-e^{i\mu\lambda}}{1-\bar{\kappa}e^{i\mu\lambda}},$$

and since we require that $\omega(i)=0$, we have

$$\omega(\lambda)=\frac{e^{\mu}-e^{-i\mu\lambda}}{1-e^{\mu}e^{-i\mu\lambda}}.$$

Comparing this with the formula (18), we see that the characteristic function of the operator A coincides with the characteristic function of the operator of differentiation on the interval $[0,\mu]$ with the boundary conditions zero at the end-points.

By Theorem 3 the operator A is unitarily equivalent to the operator of differentiation, as we had to show.

Theorem 3 prompts the question whether a symmetric operator (or an isometric operator) corresponding to a given characteristic function exists. For isometric operators the answer is affirmative; in fact we have

Theorem 5. *From any given function $w(\zeta)$ which is regular in the unit disc and maps that region into itself and which satisfies the normalization condition $w(0)=0$ an isometric operator V can be constructed which has $w(\zeta)$ as its characteristic function.*

Proof. We introduce a function $\Phi(\zeta)$ defined by

$$w(\zeta)=\frac{\Phi(\zeta)-1}{\Phi(\zeta)+1}.$$

Then, as in §69, $\Phi(\zeta)$ can be expressed as

$$\Phi(\zeta)=\int_0^{2\pi}\frac{e^{is}+\zeta}{e^{is}-\zeta}\,d\sigma(s),$$

where $\sigma(s)$ is a non-decreasing function with its total variation equal to 1.

We construct the space $L^2_\sigma(0,2\pi)$ and in it the unitary operator $\mathring{U}$ of multiplication by e^{it}:

$$\mathring{U}f(t)=e^{it}f(t).$$

Let V be the isometric operator which coincides with the operator $\mathring{U}$ on the hyperplane $D(V)$ orthogonal to the function $g(t)\equiv 1$. Then it is easy to see that V is the required operator.

For, since the resolution of the identity $\mathring{E}_t$ for the operator $\mathring{U}$ is defined by

$$\mathring{E}_t f(s) = \begin{cases} f(s) & (s \leqslant t), \\ 0 & (s > t), \end{cases}$$

we have

$$((\mathring{U} - \zeta I)^{-1} g, g) = \int_0^{2\pi} \frac{d\sigma(s)}{e^{is} - \zeta},$$

$$((\mathring{U} - \zeta I)^{-1} \mathring{U} g, g) = \int_0^{2\pi} \frac{e^{is}\, d\sigma(s)}{e^{is} - \zeta}$$

and

$$\frac{\zeta((\mathring{U} - \zeta I)^{-1} g, g)}{((\mathring{U} - \zeta I)^{-1} \mathring{U} g, g)} = \frac{\Phi(\zeta) - 1}{\Phi(\zeta) + 1} = w(\zeta),$$

and this (*cf.* formula (15)) proves the theorem.

If, for the operator V constructed from a given characteristic function $w(\zeta)$, the manifold $R(V - I)$ is not dense in H, then it is not possible to go over from V to its Cayley transform A, and so in this case there is no symmetric operator A which has

$$\omega(\lambda) = w\left(\frac{\lambda - i}{\lambda + i}\right) \tag{22}$$

as its characteristic function.

In order that the manifold $R(V - I)$ shall be dense in H, an additional restriction has to be imposed on the function (22). This matter is cleared up by

Theorem 6. *A given function $\omega(\lambda)$ which is regular in the upper half-plane, and maps that region into the interior of the unit disc, and which satisfies the normalization condition $\omega(i) = 0$ will be the characteristic function of a simple, symmetric operator with deficiency index* (1, 1) *if and only if, for any $\varepsilon > 0$ and any α $(0 \leqslant \alpha < 2\pi)$*

$$\lim_{\lambda \to \infty} \lambda\{\omega(\lambda) - e^{i\alpha}\} = \infty \qquad (0 < \varepsilon \leqslant \arg \lambda \leqslant \pi - \varepsilon).$$

Proof. We shall show that a necessary and sufficient condition for the closure of the manifold $R(V - I)$ to coincide with H is that there shall be a unitary extension, of the operator V, which has an eigenvalue equal to 1.

The sufficiency of the condition is clear, since if U is a unitary extension of the operator V, and if $U\psi = \psi$ $(\psi \neq 0)$, then, for any

$\varphi \in \mathrm{D}(V)$, we have

$$((V-I)\varphi, \psi) = ((U-I)\varphi, \psi) = (\varphi, (U^{-1}-I)\psi) = 0.$$

To prove the necessity of the condition, suppose that, for any $\varphi \in \mathrm{D}(V)$,

$$((V-I)\varphi, \psi) = 0 \qquad (\|\psi\| = 1).$$

By expressing an arbitrary vector $f \in \mathrm{H}$ in the form $f = \varphi + \gamma g$, we obtain, for an arbitrary, quasi-unitary extension S_κ of the operator V,

$$((S_\kappa - I)f, \psi) = (V\varphi - \varphi + \gamma\kappa g^* - \gamma g, \psi) = \gamma\{\kappa(g^*, \psi) - (g, \psi)\}. \qquad (23)$$

We observe that $\qquad (g'', \psi) \neq 0$
for otherwise the vector ψ could be expressed as

$$\psi = V\varphi_1 \qquad (\varphi_1 \in \mathrm{D}(V), \|\varphi_1\| = 1),$$

so that

$$((V-I)\varphi_1, V\varphi_1) = 0,$$

i.e.,

$$(\varphi_1, V\varphi_1) = 1.$$

But this equality can hold only if $V\varphi_1 = \varphi_1$, and this contradicts the fact that the operator V is simple.

Thus $\qquad (g^*, \psi) \neq 0, \qquad$ and we may therefore put

$$\kappa = \frac{(g, \psi)}{(g^*, \psi)}$$

in (23), and then, for any $f \in \mathrm{H}$, we have

$$((S_\kappa - I)f, \psi) = 0$$

or

$$(f, S_\kappa^* - I)\psi) = 0.$$

Since $S_\kappa^* = S_\eta^{-1}$, where $\eta = \bar{\kappa}^{-1}$, the last equality implies that $S_\eta^{-1}\psi = \psi$, and therefore $S_\eta \psi = \psi \quad (\psi \neq 0)$; but this is possible only if $|\eta| = |\kappa| = 1$, and so $S_\kappa = S_\eta$ is indeed the required unitary externsion, of the operator V, with an eigenvalue equal to 1.

The proof of the theorem is now reduced to finding necessary and sufficient conditions which the characteristic function of an isometric operator V must satisfy in order that none of its unitary extensions shall have an eigenvalue equal to 1.

Let $\mathring{U}$ be that unitary extension, of the operator V, which appears in formula (15), and let $U^{(\alpha)}$ be an arbitrary unitary extension of V such that $U^{(\alpha)}g = e^{i\alpha}g^*$. Since the deficiency element g is a generating element for each of the operators $U^{(\alpha)}$, the fact that the operators $U^{(\alpha)}$ $(0 \leqslant \alpha < 2\pi)$ have no eigenvalues equal to 1 is equivalent to the fact that the distribution functions $(F_t^{(\alpha)}g, g)$, where $F_t^{(\alpha)}$ is the spectral function of the operator $U^{(\alpha)}$ $(0 \leqslant \alpha < 2\pi)$, have no discontinuity at $t = 0$.

On the other hand, if $w_\alpha(\zeta)$ is the characteristic function constructed by replacing $\mathring{U}$ in (15) by $U^{(\alpha)}$, then, by the formulae (15c) and (17) we have

$$w_\alpha(\zeta) = e^{-i\alpha} w(\zeta),$$

and therefore, by (15a, b),

$$\frac{e^{i\alpha}}{e^{i\alpha} - w(\zeta)} = \int_0^{2\pi} \frac{e^{it}}{e^{it} - \zeta} \, d(F_t^{(\alpha)}g, g).$$

Hence we easily find that the magnitude σ of the discontinuity of the function $(F_t^{(\alpha)}g, g)$ at the point $t = 0$ is given by

$$\sigma = \lim_{\zeta \to 1} \frac{(1-\zeta)e^{i\alpha}}{e^{i\alpha} - w(\zeta)},$$

where $\zeta \to 1$ from inside the disc $|\zeta| < 1$ along any non-tangential direction.

Using the indications given, the proof of the theorem can easily be completed, and this is left to the reader.

§115. On triangular resolutions of certain non-self-adjoint operators

Analogous to the characteristic function of a quasi-self-adjoint extension of a symmetric operator with deficiency index (1, 1), M. S. Lipshits has constructed characteristic matrix-functions for a wide class of non-self-adjoint operators. The nature of the restrictions which have to be imposed on these operators is that they are to be, in a certain sense, little different from self-adjoint operators. Thus, for example, in the case of a bounded operator T, defined everywhere in H, it is required that its imaginary part, $(T - T^*)/(2i)$, shall be completely continuous. By means of the so-called 'multiplication theorem' for the characteristic matrix-functions (see the footnote on p. 423) the first triangular resolutions were obtained for non-self-adjoint operators in H. These expansions are the infinite-dimensional analogues of Schur's theorem on the reduction of a finite matrix to triangular form by means of unitary transformations. For some classes of non-self-adjoint operators the triangular resolution can be deduced directly from von Neumann's theorem (see §66) on the existence of a non-trivial sub-space for any completely continuous operator. In this section, following M. S. Brodsky,† we carry out the derivation for *Volterra operators, i.e.* for completely continuous operators having only the one point $\lambda = 0$ in the spectrum. This derivation uses little more than the results of Chapter 5 and the concept of the resolution of the identity (§67).

We begin with the finite-dimensional case. Let T be a Volterra operator in a Euclidean space E. Then, by Schur's theorem, there is an orthonormal basis $\{e_k\}_{k=1}^{\infty}$ for the space E such that

$$\left.\begin{aligned} &Te_1 = 0 \\ &Te_k = t_{k,1}e_1 + t_{k,2}e_2 + \cdots + t_{k,k-1}e_{k-1} \qquad (k > 1) \end{aligned}\right\}. \tag{1}$$

If P_k is the ortho-projector on to the one-dimensional sub-space containing e_k, then

$$P_kTP_j = 0 \qquad (k \geq j),$$

$$P_kT^*P_j = 0 \qquad (k \leq j),$$

and so

$$T = \sum_{j,k=1}^{n} P_kTP_j = \sum_{j=2}^{n} \sum_{1 \leq k < j} P_kTP_j = 2i \sum_{j=2}^{n} \sum_{1 \leq k < j} P_k \frac{T - T^*}{2i} P_j. \tag{2}$$

We now introduce a set $\mathcal{M}$ consisting of all the $(n+1)$ points

$$0 = \mu_0 < \mu_1 < \cdots < \mu_n = 1$$

and specify an operator-function $E(\mu)$ on $\mathcal{M}$ by putting

$$E(0) = 0, \qquad E(\mu_k) = P_1 + P_2 + \cdots + P_k \qquad (k = 1, 2, \ldots, n).$$

Then the equality (2) can be put in the form

$$T = 2i \int_{\mathcal{M}} E(\mu) K \, dE(\mu), \tag{3}$$

where

$$K = \frac{1}{2i}(T - T^*). \tag{4}$$

The purpose of the rest of this section is to generalize the triangular spectral expansion (3) to arbitrary Volterra operators in a separable Hilbert space H. In doing so we shall use the following definition of an integral of operator functions.

Let $\mathcal{M}$ be any closed set of points, of the interval [0, 1], which includes

† See Brodsky [19], pp. 135–141. A more complete account and further developments are to be found in Gokhberg and Krein [40].

the end-points. We shall call the set of numbers

$$0=\mu_0<\mu_1<\cdots<\mu_m=1$$

belonging to $\mathcal{M}$ a *δ-partition of the set* $\mathcal{M}$ if the points μ_{k-1} and μ_k $(k=1,2,\ldots,m)$ either are the ends of an interval complementary to $\mathcal{M}$ or satisfy the condition $\mu_k-\mu_{k-1}<\delta$.

If operator functions $F(\mu)$ and $G(\mu)$ are defined on $\mathcal{M}$, then by *the integral*

$$\int_{\mathcal{M}} F(\mu)\, dG(\mu)$$

we shall mean the limit as $\delta\to 0$ of the integral sums

$$\sum_{k=1}^{m} F(\xi_k)\{G(\mu_k)-G(\mu_{k-1})\} \qquad (\mu_{k-1}\leqslant\xi_k\leqslant\mu_k,\ \xi_k\in\mathcal{M})$$

in the sense of the operator norm, if this limit exists.

Next we specialize slightly the concept of an orthogonal resolution of the identity $E(\mu)$ by relating it not to the whole interval $[0, 1]$ but just to the set $\mathcal{M}$. Thus we shall regard the function $E(\mu)$ as being defined only on $\mathcal{M}$, so that

$$E(\mu')<E(\mu'') \text{ if } \mu'<\mu'' \qquad (\mu',\mu''\in\mathcal{M}), \qquad E(0)=0, \qquad E(1)=I.$$

Further, we shall suppose that the function $E(\mu)$ is continuous on $\mathcal{M}$, and that it also satisfies the following

Condition A: If (α,β) is a complementary interval of the set $\mathcal{M}$, then $E(\beta)-E(\alpha)$ is a one-dimensional operator.

If an operator function $E(\mu)$ which has all the properties mentioned is such that, for any $\mu\in\mathcal{M}$, the sub-set $E(\mu)$H is invariant under a Volterra operator T, then we shall call $E(\mu)$ the *spectral function* of this operator. We must draw attention to the fact that μ here is not a spectral variable, as it was in the case of orthogonal spectral functions of self-adjoint operators. The spectrum of a Volterra operator consists of only a single point——$\lambda=0$——and the variable μ plays the same rôle as the index k in formula (1).

We shall prove that *any Volterra operator T in* H *has a spectral function.*

A set $\mathcal{G}$ of invariant sub-spaces G of the operator T will be called a *chain* if the following conditions hold:

1) if $G_1\in\mathcal{G}$ and $G_2\in\mathcal{G}$, then either $G_1\subset G_2$ or $G_2\subset G_1$;
2) the null sub-space and the whole of H belong to $\mathcal{G}$.

The set $\Gamma\{T\}$ of all chains belonging to the operator T can be partially ordered by taking $\mathscr{G}' \prec \mathscr{G}''$ to hold if every invariant sub-space G in $\mathscr{G}'$ also belongs to $\mathscr{G}''$. A chain $\mathscr{G}$ is said to be *maximal* if it comes after every other chain in $\Gamma\{T\}$ comparable with it. If γ is a monotone increasing sequence of chains belonging to $\Gamma\{T\}$, then the set-theoretic sum of all the invariant sub-spaces belonging to these chains is evidently maximal relative to γ, and it too belongs to $\Gamma\{T\}$. By Zorn's lemma, each chain in $\Gamma\{T\}$ is contained in some maximal chain belonging to $\Gamma\{T\}$.

Now let $\mathscr{G}$ be a maximal chain in $\Gamma\{T\}$ and let $\{g_k\}_{k=1}^{\infty}$ be some orthonormal basis in H. Let F be a sub-space of H, and let P be the ortho-projector on to F; then the number

$$\mu = \sum_{k=1}^{\infty} \frac{1}{2^k} \|Pg_k\|^2 \tag{5}$$

is called the *weight* of the sub-space F. Clearly, different sub-spaces of H which belong to the same chain cannot have the same weight. We now let $\mathscr{M}$ denote the set of weights of all the sub-spaces of some maximal chain $\mathscr{G}$, and let $E(\mu)$ be the ortho-projector on to that sub-space in $\mathscr{G}$ whose weight is μ. Then it is easily verified that the set $\mathscr{M}$ is closed, and that the operator function $E(\mu)$ has all the properties of a spectral function of the operator T, except perhaps that the condition A does not hold. The verification of this last condition is based on von Neumann's theorem in §65, and proceeds thus.

Suppose that the condition A does not hold. This means that, for some complementary interval (α, β) of the set $\mathscr{M}$, we shall have

$$\dim\{G_\beta \ominus G_\alpha\} > 1,$$

where G_α and G_β are the sub-spaces on to which $E(\alpha)$ and $E(\beta)$ project. By Theorem 1 in §65 the operator PTP, where $P = E(\beta) - E(\alpha)$, has in $G_\beta \ominus G_\alpha$ an invariant sub-space F $\quad (F \neq \{0\}, F \neq G_\beta \ominus G_\alpha)$. But then the sub-space $G_\alpha \oplus F$ is invariant under T, and is strictly included between G_α and G_β, and this contradicts the fact that the chain $\mathscr{G}$ is maximal. Thus $E(\mu)$ is a spectral function of the operator T.

We shall call the maximal chain $\mathscr{G} \in \Gamma\{T\}$ for which the weights of the sub-spaces making up $\mathscr{G}$ were introduced in accordance with formula (5) and for which the set $\mathscr{M}$ was constructed and the spectral function $E(\mu) \quad (\mu \in \mathscr{M})$ was defined the *normalized maximal chain*.

Lemma. *For any complementary interval (α, β) of the set $\mathscr{M}$ constructed for a normalized maximal chain $\mathscr{G} \in \Gamma\{T\}$,*

$$[E(\beta) - E(\alpha)]T[E(\beta) - E(\alpha)] = 0. \tag{6}$$

Proof. The sub-space G_α is invariant under T_β, the restriction of the operator T to G_β. Therefore the 1-dimensional sub-space $G_\beta \ominus G_\alpha$ is invariant under T^*_β. But since, if T is a Volterra operator, T_β and therefore also T^*_β are Volterra operators, it follows that $G_\beta \ominus G_\alpha$ is the null space of the operator T^*_β. Therefore the manifold $R(T_\beta)$ of the values of the operator T_β is orthogonal to $G_\beta \ominus G_\alpha$, and from this the relation (6) follows.

We can now proceed to M. S. Brodsky's theorem, which is the final objective of this section.

Theorem. *Any Volterra operator T admits a triangular representation*

$$T = 2i\int_{\mathcal{M}} E(\mu)K\,dE(\mu) \tag{7}$$

where K is the imaginary part, (4), *of T, and* $E(\mu)$ *is an arbitrary spectral function for T. The integral sums for the integral in* (7) *converge uniformly to the operator T.*

Proof. Let $0 = \mu_0 < \mu_1 < \cdots < \mu_m = 1$ be a δ-partition of the set $\mathcal{M}$. Writing

$$\Delta E(\mu_k) = E(\mu_k) - E(\mu_{k-1}),$$

and noting that $\Delta E(\mu_k) T \Delta E(\mu_k) = 0 \quad \text{for } k > j,$

we obtain

$$\begin{aligned} T &= \sum_{j,k=1}^{m} \Delta E(\mu_k) T \Delta E(\mu_j) = \sum_{j=1}^{m} \sum_{1 \le k \le j} \Delta E(\mu_k) T \Delta E(\mu_j) \\ &= \sum_{k=1}^{m} \Delta E(\mu_k) T \Delta E(\mu_k) + 2i \sum_{j=2}^{m} \sum_{1 \le k < j} \Delta E(\mu_k) K \Delta E(\mu_j) \\ &= \sum_{k=1}^{m} \Delta E(\mu_k) T \Delta E(\mu_k) + 2i \sum_{j=1}^{m} E(\mu_{j-1}) K \Delta E(\mu_j). \end{aligned}$$

Therefore, if $\mu_{j-1} \le \xi_j \le \mu_j$, we have

$$\begin{aligned} &\left\| T - 2i \sum_{j=1}^{m} E(\xi_j) K \Delta E(\mu_j) \right\| \\ &\quad \le \left\| T - 2i \sum_{j=1}^{m} E(\mu_{j-1}) K \Delta E(\mu_j) \right\| \\ &\qquad + 2 \left\| \sum_{j=1}^{m} E(\mu_{j-1}) K \Delta E(\mu_j) - \sum_{j=1}^{m} E(\xi_j) K \Delta E(\mu_j) \right\| \\ &\quad \le \left\| \sum_{k=1}^{m} \Delta E(\mu_k) T \Delta E(\mu_k) \right\| + 2 \left\| \sum_{j=1}^{m} \Delta E(\mu_j) K \Delta E(\mu_j) \right\| \\ &\quad \le 3 \left\| \sum_{k=1}^{m} \Delta E(\mu_k) T \Delta E(\mu_k) \right\|. \end{aligned}$$

It now remains to prove that the norm of the operator

$$S_\delta = \sum_{k=1}^{m} \Delta E(\mu_k) T \Delta E(\mu_k) \tag{8}$$

tends to 0 as $\delta \to 0$.

We consider the operators $(1/2i)(S_\delta - S_\delta^*)$ and $\frac{1}{2}(S_\delta + S_\delta^*)$ separately. Let $\{e_j\}_{j=1}^{\infty}$ be an orthonormal basis of the eigenvectors of the operator K in its range, and let $\{\lambda_j\}_{j=1}^{\infty}$ be the corresponding sequence of eigenvalues. Then, for any $f \in \mathrm{H}$,

$$\begin{aligned}
\left|\left(\frac{1}{2i}(S_\delta - S_\delta^*)f, f\right)\right| &= \left|\sum_{k=1}^{m} (\Delta E(\mu_k) K \Delta E(\mu_k) f, f)\right| \\
&\leqslant \sum_{k=1}^{m} \sum_{j=1}^{\infty} |(\Delta E(\mu_k) f, e_j)|^2 \, |\lambda_j| \\
&\leqslant \sum_{k=1}^{m} \sum_{j=1}^{n} \|\Delta E(\mu_k) f\|^2 \, \|\Delta E(\mu_k) e_j\|^2 \, |\lambda_j| \\
&\quad + \sum_{k=1}^{m} \sum_{j=n+1}^{\infty} |(\Delta E(\mu_k) f, e_j)|^2 \, |\lambda_j| \\
&\leqslant \|f\|^2 \left\{ \sum_{j=1}^{n} |\lambda_j| \max_{\substack{1 \leqslant k \leqslant m \\ 1 \leqslant j \leqslant n}} \|\Delta E(\mu_k) e_j\|^2 + \max_{j > n} |\lambda_j| \right\}.
\end{aligned}$$

The second term in the braces can be made as small as we please by choosing n suitably. In the first term we need consider only those values of the subscript k for which $\mu_k - \mu_{k-1} < \delta$, because, by (6), for other values of k, the corresponding terms in (8) disappear. So, using the uniform continuity of the functions $E(\mu)e_j$ on the closed set $\mathcal{M}$, the first term in the braces can also be made arbitrarily small by letting δ decrease, for a fixed n. Thus

$$\lim_{\delta \to 0} \|S_\delta - S_\delta^*\| = 0,$$

and similarly

$$\lim_{\delta \to 0} \|S_\delta + S_\delta^*\| = 0.$$

Hence the theorem.

Appendix 1

Integral operators

§116. Definitions and auxiliary results

Among linear integral operators the simplest are the Hilbert–Schmidt operators (see §32). The kernel $K(s, t)$ of such an operator satisfies the inequality

$$\iint_{-\infty}^{\infty} |K(s, t)|^2 \, ds \, dt < \infty .$$

In the classical papers by D. Hilbert (see [44]) the more general case is studied, where, for any function $f(t) \in L^2(-\infty, \infty)$,

$$\int_{-\infty}^{\infty} ds \left| \int_{-\infty}^{\infty} K(s, t) f(t) \, dt \right|^2 \leq M^2 \int_{-\infty}^{\infty} |f(t)|^2 \, dt,$$

M being a constant. In this case, the kernel $K(s, t)$ generates a bounded integral operator.

An even more general case occurs if it is assumed that, for almost all $s \in (-\infty, \infty)$,

$$\int_{-\infty}^{\infty} |K(s, t)|^2 \, dt < \infty .$$

The study of integral operators with a symmetric kernel satisfying this inequality is the subject of this Appendix. The theory of such operators and also of a more general kind of integral operators is due to Carleman.†

From now on we shall be concerned with the space L^2 of functions on the whole real line, although the theory is still true for functions of several variables, and also in the case where, instead of the whole of a

† This theory was developed during the years 1920–1923. The first systematic exposition was given by T. Carleman. M. Stone gave a rather different treatment. The present account is based on an article by N. I. Akhiezer. See Carleman [22], Stone [119], and Akhiezer [3].

(one-dimensional or multi-dimensional) Euclidean space, a point-set of positive Lebesgue measure in that space is chosen.

Definition. A *Carleman kernel* is a measurable (complex-valued) function

$$K(s, t) \qquad (-\infty < s, t < \infty) \quad \text{such that}$$

a) almost everywhere in the s,t-plane

$$\overline{K(s, t)} = K(t, s),$$

b) almost everywhere on the s-axis

$$\int_{-\infty}^{\infty} |K(s, t)|^2 \, dt < \infty.$$

We put

$$K(s) = \sqrt{\left\{\int_{-\infty}^{\infty} |K(s, t)|^2 \, dt\right\}} \tag{1}$$

if the right-hand side is finite, and otherwise put $K(s) = 0$.

We observe first of all that, almost everywhere,

$$\int_{-\infty}^{\infty} |K(t, s)|^2 \, dt = K^2(s). \tag{2}$$

For, by condition a) we have

$$\iint_{-\infty}^{\infty} |K(s, t) - \overline{K(t, s)}|^2 \, ds \, dt = 0,$$

and hence it follows by Fubini's theorem that, for almost all s,

$$\int_{-\infty}^{\infty} |K(s, t) - \overline{K(t, s)}|^2 \, dt = 0.$$

But this implies that, for almost all s,

$$\int_{-\infty}^{\infty} |K(t, s)|^2 \, dt = \int_{-\infty}^{\infty} |K(s, t)|^2 \, dt,$$

and (2) has been proved.

We now introduce integral operators generated by Carleman kernels. If $f \in L^2$, then

$$h(s) = \int_{-\infty}^{\infty} K(s, t) f(t) \, dt \tag{3}$$

exists for almost all s. However, $h(s)$ may not belong to L^2. We therefore distinguish first of all the set D^* of all functions $f(t) \in L^2$ such that $h(s) \in L^2$. On this set D^* the formula (3) defines a linear operator, which we denote by A^*. Every other integral operator in L^2 which has this kernel $K(s, t)$ is a restriction of A^*.

We next define one such operator. First let us agree on a bit of notation: if $P(s) \quad (-\infty < s < \infty)$ is a fixed, measurable, real-valued, non-negative function, then we shall denote by $[L^2]_P$ the set of all functions $\varphi(s) \in L^2(-\infty, \infty)$ such that

$$\int_{-\infty}^{\infty} P(s)\, |\varphi(s)|\, ds < \infty,$$

and we shall denote the left-hand side of this inequality by $\|\varphi\|_P$.

We now show that $[L^2]_K \subseteq D^*$, where $K(s)$ is defined by (1) almost everywhere. For, let $f(t) \in [L^2]_K$, *i.e.*,

$$f(t) \in L^2 \quad \text{and} \quad \int_{-\infty}^{\infty} K(s)\, |f(s)|\, ds < \infty.$$

We have to show that in that case

$$\int_{-\infty}^{\infty} \left| \int_{-\infty}^{\infty} K(s, t) f(t)\, dt \right|^2 ds < \infty.$$

But this follows from Fubini's theorem, because

$$\begin{aligned} &\int_{-\infty}^{\infty} |K(s, u)\overline{K(s, v)} f(u)\overline{f(v)}|\, ds \\ &\quad \leq |f(u)| \,.\, |f(v)| \,.\, \sqrt{\left\{\int_{-\infty}^{\infty} |K(s, u)|^2\, ds\right\}} \,.\, \sqrt{\left\{\int_{-\infty}^{\infty} |K(s, v)|^2\, ds\right\}} \qquad (4) \\ &\quad = K(u)K(v)\, |f(u)| \,.\, |f(v)| \end{aligned}$$

almost everywhere in the u, v-plane; and, on the other hand,

$$\iint_{-\infty}^{\infty} K(u)K(v)\, |f(u)| \,.\, |f(v)|\, du\, dv = \left\{\int_{-\infty}^{\infty} K(u)\, |f(u)|\, du\right\}^2 < \infty.$$

We now define an operator A_0 with the domain $D(A_0) = [L^2]_K$ by putting

$$A_0 f = A^* f \qquad (f \in D(A_0)).$$

We note that the domain $D(A_0)$ is dense in L^2. The truth of this assertion follows from the next lemma.

We also note that, from (4),

$$\|A_0 f\| \leq \|f\|_K. \qquad (5)$$

Lemma. *If the measurable function $P(s) \geqslant 0$ $(-\infty < s < \infty)$ is almost everywhere finite, and if the function $h(s)$ is measurable and such that, for any function $f(s) \in [L^2]_P$,*

$$\int_{-\infty}^{\infty} f(s)h(s)\,ds = 0,$$

then $h(s) = 0$ *almost everywhere.* (6)

Proof. First we observe that $h(s)$ can become infinite only on a set of measure zero. Further, we let $e(a)$ denote the set of all s such that

$$P(s) \leqslant a, \qquad |h(s)| \leqslant a,$$

and let e be an arbitrary sub-set, of finite measure, of the set $e(a)$. Its characteristic function $\chi_e(s)$ clearly belongs to $[L^2]_P$. Therefore

$$\int_e h(s)\,ds = \int_{-\infty}^{\infty} \chi_e(s)h(s)\,ds = 0.$$

Since e is arbitrary, the equality (6) holds almost everywhere in $e(a)$, and then, since a is arbitrary, it holds almost everywhere on the whole axis.

Q.E.D.

Corollary. *The set $[L^2]_P$ is dense in L^2.*

Theorem. *The operator A_0 is symmetric, and the operator A^* is adjoint to A_0.*

Proof. Let $f(t) \in D(A_0)$ and $g(t) \in L^2$; then almost everywhere we have

$$\int_{-\infty}^{\infty} |K(s,t)| \,.\, |f(t)\overline{g(s)}|\,ds$$

$$\leqslant |f| \,.\, \sqrt{\left\{\int_{-\infty}^{\infty} |K(s,t)|^2\,ds\right\}} \,.\, \sqrt{\left\{\int_{-\infty}^{\infty} |g(s)|^2\,ds\right\}} = K(t)\,|f(t)| \,.\, \|g\|.$$

Therefore

$$\iint_{-\infty}^{\infty} |K(s,t)| \,.\, |f(t)\overline{g(s)}|\,ds\,dt \leqslant \|f\|_K \,.\, \|g\|,$$

which implies that

$$\int_{-\infty}^{\infty} \left\{\int_{-\infty}^{\infty} K(s,t)f(t)\,dt\right\}\overline{g(s)}\,ds = \int_{-\infty}^{\infty} f(t)\,dt \int_{-\infty}^{\infty} \overline{K(t,s)g(s)}\,ds. \qquad (7)$$

If we assume that $g(t)$, as well as $f(t)$, belongs to $D(A_0)$, then this relation can be written as

$$(A_0 f, g) = (f, A_0 g),$$

and so the first part of the theorem is proved.

To prove the second assertion, we observe first of all that, if $f \in D(A_0)$ and $g \in D^*$, then the relation (7) can be expressed as

$$(A_0 f, g) = (f, g^*),$$

where

$$g^*(t) = \int_{-\infty}^{\infty} K(t, s) g(s)\, ds$$

is some element in the space L^2. Hence it follows that $g(t) \in D(A_0^*)$, and thus we have proved that $D^* \subseteq D(A_0^*)$.

We now have to prove the reverse inclusion. To do so, let us assume that $g(t) \in D(A_0^*)$. This implies that there is a function $g^*(t) \in L^2$ such that, for any $f \in D(A_0)$,

$$\int_{-\infty}^{\infty} \left\{ \int_{-\infty}^{\infty} K(s, t) f(t)\, dt \right\} \overline{g(s)}\, ds = \int_{-\infty}^{\infty} f(t) \overline{g^*(t)}\, dt.$$

By (7) this can be re-written as

$$\int_{-\infty}^{\infty} f(t)\, dt \left\{ \int_{-\infty}^{\infty} \overline{K(t, s) g(s)}\, ds - \overline{g^*(t)} \right\} = 0.$$

Hence, by the lemma, it follows that, almost everywhere

$$g^*(t) = \int_{-\infty}^{\infty} K(t, s) g(s)\, ds,$$

and since the function $g^*(t)$ belongs to L^2, so $g(s)$ belongs to D^*, and $A^* g = g^*(t)$. Thus we have shown that $D^* \supseteq D(A_0^*)$. Therefore $D^* = D(A_0^*)$. Q.E.D.

Definition. The operator $A = A_0^{**}$, *i.e.*, the closure of the operator A_0, is called the *Carleman integral operator.* We shall keep the notation D^* for $D(A^*)$, and also write D instead of $D(A)$. Following Carleman, we shall say that $K(s, t)$ is a *kernel of the first kind* if the operator A is self-adjoint, and if not then $K(s, t)$ is a *kernel of the second kind.*

§117. Example

Let $\{\psi_n(s)\}_0^\infty$ be an orthonormal basis of Haar functions for the interval $[0, 1]$; these functions are defined as follows:

$$\psi_0(s)=1 \qquad (0 \leqslant s \leqslant 1)$$

$$\psi_1(s)=\begin{cases}-1 & (0\leqslant s<\tfrac{1}{2})\\ 1 & (\tfrac{1}{2}\leqslant s\leqslant 1)\end{cases}$$

$$\psi_n(s)=\begin{cases}0 & (0\leqslant s<1-1/2^{n-1})\\ -2^{(n-1)/2} & (1-1/2^{n-1}\leqslant s<1-1/2^n)\\ 2^{(n-1)/2} & (1-1/2^n\leqslant s\leqslant 1).\end{cases} \qquad (n=2, 3, \ldots).$$

Since, for every $s\in[0, 1]$, there is a p_s such that $\psi_p(s)=0$ when $p\geqslant p_s$, the series

$$\sum_{p=0}^{\infty} c_p\psi_p(s) \quad \text{and} \quad \sum_{p=0}^{\infty} c_p\psi_p^2(s)$$

converge absolutely in the interval $0\leqslant s<1$, whatever the coefficients c_p.

Lemma 1. *The sum*

$$\sum_{p=0}^{\infty} c_p\psi_p(s)=\theta(s)$$

is equal to 0 *almost everywhere in the interval* $0\leqslant s<1$ *if and only if*

$$c_p=2^{(p-1)/2}c_0 \qquad (p=1, 2, 3, \ldots).$$

Proof. We express the interval $0\leqslant s<1$ as the sum of the intervals

$$(d_n): \qquad 1-1/2^{n-1}\leqslant s<1-1/2^n \qquad (n=1, 2, 3, \ldots).$$

Since, in the interval (d_1),

$$\theta(s)=c_0-c_1,$$

and in the interval (d_n) $\quad (n\geqslant 2)$,

$$\theta(s)=c_0+c_1+2^{1/2}c_2+2^{2/2}c_3+\cdots+2^{(n-2)/2}c_{n-1}-2^{(n-1)/2}c_n,$$

in order that the function $\theta(s)$ shall be equal to 0 at at least one point in each of the intervals (d_n) $\quad (n=1, 2, 3, \ldots)$, all the equalities

$$c_0-c_1=0$$

$$c_0+c_1+2^{1/2}c_2+2^{2/2}c_3+\cdots+2^{(n-2)/2}c_{n-1}-2^{(n-1)/2}c_n=0 \qquad (n=2, 3, 4, \ldots)$$

must hold.

If all these equalities hold, then $\theta(s)=0$ everywhere in the interval $0 \leqslant s < 1$, and from this the assertion of the lemma follows.

Below we shall need the functions

$$\kappa_1(s)=\psi_0(s)-\psi_1(s),$$

$$\kappa_n(s)=\psi_0(s)+\psi_1(s)+2^{1/2}\psi_2(s)+\cdots+2^{(n-2)/2}\psi_{n-1}(s)-2^{(n-1)/2}\psi_n(s)$$
$$(n=2,3,\ldots).$$

By using the proof of the Lemma 1 it is easily seen that

$$\kappa_1(s)=\begin{cases}2 & (0\leqslant s<1/2),\\ 0 & (1/2\leqslant s<1),\end{cases}$$

$$\kappa_n(s)=\begin{cases}0 & (0\leqslant s<1-1/2^{n-1}),\\ 2^n & (1-1/2^{n-1}\leqslant s<1-1/2^n),\\ 0 & (1-1/2^n\leqslant s<1).\end{cases} \qquad (n=2,3,\ldots)$$

Thus the function $\kappa_n(s)$ $\quad (n=1,2,3,\ldots)$ has a constant, non-zero value in the interval (d_n) and is equal to 0 outside this interval.

Lemma 2. *Every function $\varphi(s)\in L^2(0,1)$ can be expressed as a convergent series*

$$\varphi(s)=\sum_{p=0}^{\infty} c_p\psi_p(s)+\omega(s),$$

where

$$\omega(s)\in L^2(0,1), \qquad \sum_{p=0}^{\infty}|c_p|^2<\infty,$$

$$c_p=\int_0^1 \varphi(s)\psi_p(s)\,ds, \qquad \int_0^1 \omega(s)\psi_p(s)=0 \qquad (p=0,1,2,\ldots).$$

Proof. The sequence of partial sums

$$\theta_n(s)=\sum_{p=0}^{n} c_p\psi_p(s)$$

converges in the metric of $L^2(0,1)$ to some function $\theta(s)\in L^2$. Therefore (see §11) there is a sub-sequence $\{\theta_{n_k}(s)\}_{k=1}^{\infty}$ which converges to $\theta(s)$ almost everywhere. But since the series

$$\sum_{p=0}^{\infty} c_p\psi_p(s) \qquad (0\leqslant s<1)$$

converges everywhere, its sum is equal to $\theta(s)$ almost everywhere, and to prove the lemma it only remains to observe that

$$\int_0^1 \theta(s)\psi_p(s)\,ds = c_p \qquad (p = 0, 1, 2, \ldots). \qquad \text{Q.E.D.}$$

We now consider the function

$$K(s, t) = \sum_{p=0}^{\infty} a_p \psi_p(s)\psi_p(t) \qquad (0 \leq s, t < 1), \tag{1}$$

where the a_p are real numbers. It is easily seen that this is a Carleman kernel, and that

$$K^2(s) = \int_0^1 |K(s, t)|^2\,dt = \sum_{p=0}^{\infty} a_p^2 \psi_p^2(s). \tag{2}$$

The properties of the kernel $K(s, t)$ will, of course, depend on the way in which the numbers a_p are chosen.

We show, firstly, that $K(s, t)$ will be a Hilbert–Schmidt kernel if and only if the series

$$\sum_{p=0}^{\infty} a_p^2 \tag{3}$$

converges.

For, by the theorem on termwise integration of a series with non-negative terms, we have, because of (2),

$$\int_0^1 \int_0^1 |K(s, t)|^2\,ds\,dt = \sum_{p=0}^{\infty} a_p^2,$$

where the left-hand side and the right-hand side are either both finite or both infinite, and from this our assertion follows.

Since we are not concerned with Hilbert–Schmidt kernels here, we shall assume that the series (3) diverges, and we shall consider the following alternatives:

$$(\mathscr{D}) \qquad \sum_{p=0}^{\infty} \frac{2^{p-1}}{1 + a_p^2} = \infty$$

$$(\mathscr{C}) \qquad \sum_{p=0}^{\infty} \frac{2^{p-1}}{1 + a_p^2} < \infty.$$

We shall prove that in the case $(\mathscr{D})$, the kernel $K(s, t)$ is of the first kind, and in case $(\mathscr{C})$ a kernel of the second kind.

Which kind the kernel is depends on the deficiency index of the operator A generated by the kernel $K(s, t)$. The deficiency numbers are equal here, because $K(s, t)$ is real-valued. To determine them we have to find the eigenfunctions, of the operator A^*, which correspond to an arbitrarily chosen non-real number λ, to $\lambda = i$ say. By Lemma 2, the required eigenfunction can be expressed in the form

$$\varphi(s) = \sum_{p=0}^{\infty} c_p \psi_p(s) + \omega(s),$$

where the numbers c_p and the function $\omega(s)$ must satisfy all the conditions of Lemma 2. Moreover, the function $\varphi(s)$ must satisfy the equation

$$A^* \varphi = i\varphi. \tag{4}$$

We select the interval (d_n); in it the left-hand side of (4) is equal almost everywhere to

$$A^* \varphi = \sum_{p=0}^{n} a_p \psi_p(s) \int_0^1 \varphi(t) \psi_p(t)\, dt = \sum_{p=0}^{n} a_p c_p \psi_p(s),$$

and the right-hand side is equal to

$$i \left\{ \sum_{p=0}^{n} c_p \psi_p(s) + \omega(s) \right\}.$$

Thus

$$\omega(s) = -\sum_{p=0}^{n} (1 + i a_p) c_p \psi_p(s)$$

almost everywhere in the interval (d_n), and this equality shows that $\omega(s)$ is constant in each interval (d_n). Let

$$\omega(s) - \gamma_n \qquad (s \in (d_n)) \qquad (n = 1, 2, 3, \ldots).$$

Since $\omega(s)$ is orthogonal to all the functions $\psi_p(s)$, so, remembering the properties of the functions $\kappa_n(s)$, we find that

$$0 = \int_0^1 \omega(s) \kappa_n(s)\, ds = \int_{(d_n)} \omega(s) \kappa_n(s)\, ds = \gamma_n 2^n \,.\, 1/2^n = \gamma_n$$
$$(n = 1, 2, 3, \ldots).$$

As we see, the function $\omega(s)$ must equal 0, and the coefficients c_p must be such that

$$\sum_{p=0}^{\infty} (1 + i a_p) c_p \psi_p(s) = 0\dagger$$

almost everywhere in the interval $0 \leq s < 1$. By Lemma 1, this equality

† We note that the series on the left-hand side does not converge in $L^2(0, 1)$.

holds if and only if

$$(1+ia_p)c_p = 2^{(p-1)/2}(1+ia_0)c_0 \qquad (p=1, 2, 3, \ldots),$$

i.e., if

$$|c_p|^2 = |c_0|^2(1+a_0^2)\frac{2^{p-1}}{1+a_p^2} \qquad (p=1, 2, 3, \ldots).$$

The formulae obtained show that in case $(\mathscr{D})$ the point i is not an eigenvalue of the operator A^*, and in case $(\mathscr{C})$ it is an eigenvalue of A^* —— with multiplicity 1. Our assertion is thus proved, and we have also shown that in case $(\mathscr{C})$ the deficiency index of the operator A is $(1, 1)$.

§ 118. Spectral functions of an integral operator with a Carleman kernel

Let A be an integral operator with a Carleman kernel $K(s, t)$, and let $E_\lambda \quad (-\infty \leqslant \lambda \leqslant \infty)$ be any one of the spectral functions belonging to this operator. We put

$$F_\lambda = \begin{cases} E_\lambda & (\lambda < 0), \\ 0 & (\lambda = 0), \\ E_\lambda - I & (\lambda > 0), \end{cases}$$

so that $F_\lambda \operatorname{sgn} \lambda$ is a negative, self-adjoint operator with a lower bound $\geqslant -1$. From these properties of the operator F_λ it follows that the operator $F_\lambda^2 + F_\lambda \operatorname{sgn} \lambda$ is negative. Therefore, for any $f \in \mathrm{L}^2$, we have

$$(F_\lambda f, F_\lambda f) \leqslant |(F_\lambda f, f)|. \tag{1}$$

We shall also need the inequality

$$|(F_\lambda f, f)| \leqslant \frac{1}{\lambda^2}\|f\|_K^2 \tag{2}$$

which holds for any $f \in [\mathrm{L}^2]_K$. We shall prove the result for $\lambda > 0$. To do this, it suffices to observe that, for any $f \in [L^2]_K = \mathrm{D}(A_0)$,

$$(A_0 f, A_0 f) = (Af, Af) = \int_{-\infty}^{\infty} \mu^2 d(E_\mu f, f)$$

$$\geqslant \lambda^2 \int_\lambda^\infty d(E_\mu f, f) = \lambda^2((I - E_\lambda)f, f) = \lambda^2 |(F_\lambda f, f)|,$$

and on the other hand (see (4) in §116),

$$(A_0 f, A_0 f) \leqslant \|f\|_K^2.$$

The proof for $\lambda < 0$ is similar.

Theorem 1. *F_λ is an integral operator with a Carleman kernel $F(t, s; \lambda)$ which, for every non-zero λ satisfies the following inequality almost everywhere on the t-axis:*

$$\int_{-\infty}^{\infty} |F(t, s; \lambda)|^2\, ds \leq \frac{K_1^2(t)}{\lambda^2}, \tag{3}$$

where

$$K_1(t) = \max\{1, K(t)\}.$$

Proof. For convenience we divide the proof into 4 stages.

1°. First of all we observe that, by Lemma 1 of §25, there is a function $G(s, t; \lambda)$ which, as a function of t, belongs to L^2 for each fixed s, and which satisfies $G(0, t; \lambda) = 0$, and which is such that, for any function $f(t) \in L^2$,

$$(F_\lambda f)(s) = \frac{d}{ds}\int_{-\infty}^{\infty} G(s, x; \lambda) f(x)\, dx \tag{4}$$

for almost all† s.

We shall prove that, for each s, we can change the function $G(s, t; \lambda)$ on some set of measure zero on the t-axis so that the changed function will, for each t, have almost everywhere a derivative with respect to s

$$F(s, t; \lambda) = \frac{\partial}{\partial s} G(s, t; \lambda)$$

which belongs to L^2 (as a function of s).

To do this, we put

$$k(s) = \begin{cases} 1/K(s) & \text{if } K(s) \geq 1, \\ 1 & \text{if } K(s) < 1, \end{cases}$$

and apply formula (4) to the function

$$f_{t,h}(x) = \begin{cases} \dfrac{1}{h} k(x) & \text{if } x \in (t, t+h), \\ 0 & \text{if } x \notin (t, t+h), \end{cases}$$

which, for any $h > 0$, belong to $[L]_K$. We find that, for almost all s,

$$(F_\lambda f_{t,h})(s) = \frac{d}{ds}\frac{1}{h}\int_t^{t+h} G(s, x; \lambda) k(x)\, dx = \frac{d}{ds} G_h(s, t; \lambda),$$

† The set of measure zero on which (4) does not hold depends on $f(t)$ and also on λ. This fact must be borne in mind later. Incidentally, in the present considerations, the non-zero value of λ can be regarded as fixed.

where

$$G_h(s, t; \lambda) = \frac{1}{h}\int_t^{t+h} G(s, x; \lambda)k(x)\,dx. \tag{5}$$

Therefore, by (1) and (2),

$$\int_{-\infty}^{\infty} \left|\frac{d}{ds} G(s, t; \lambda)\right|^2 ds = (F_\lambda f_{t,h}, F_\lambda f_{t,h}) \leqslant \frac{1}{\lambda^2}\|f_{t,h}\|_K^2$$

$$= \frac{1}{\lambda^2}\left\{\frac{1}{h}\int_t^{t+h} K(s)k(s)\,ds\right\}^2 \leqslant \frac{1}{\lambda^2}. \tag{6}$$

Hence if follows that there is a sequence $\{h_n\}$ dependent on t and such that $h_n \to 0$ $(h_n > 0)$ and that there are functions (of s) $g(s, t; \lambda)$ belonging to L^2 and such that, as $n \to \infty$,

$$\frac{d}{ds} G_{h_n}(s, t; \lambda) \xrightarrow{\text{weakly}} g(s, t; \lambda).$$

This implies that, for each s,

$$\lim_{n\to\infty} G_{h_n}(s, t; \lambda) = \int_0^s g(x, t; \lambda)\,dx.$$

On the other hand, by (5), for each s we have, for almost all t,

$$\lim_{h\to 0} G_h(s, t; \lambda) = G(s, t; \lambda)k(t).$$

Therefore, for each s,

$$G(s, t; \lambda)k(t) = \int_0^s g(x, t; \lambda)\,dx \qquad \text{for almost all } t. \tag{7}$$

By changing the function $G(s, t; \lambda)$, for each s, on some set of measure zero on the t-axis, we can arrange that (7) shall hold for all s and all t. Since $k(t) \neq 0$, the changed function $G(s, t; \lambda)$ will, for each t, and for almost all s, have a derivative

$$F(s, t; \lambda) = \frac{\partial}{\partial s} G(s, t; \lambda),$$

which, as a function of s, will belong to L^2.

2°. The next step consists in proving that, for every function $f \in L^2$, the equality

$$(F_\lambda f)(t) = \int_{-\infty}^{\infty} \overline{F(s, t; \lambda)} f(s)\,ds \tag{8}$$

holds almost everywhere on the t-axis. This fact is very easily proved for an elementary step-function (in a finite interval $[\alpha, \beta]$):

$$\chi_{\alpha\beta}(s) = \begin{cases} 1 & \text{if } s \in [\alpha, \beta), \\ 0 & \text{if } s \notin [\alpha, \beta), \end{cases}$$

and therefore also for any step-function with compact support. Because, for any function $h(t) \in L^2$,

$$\int_{-\infty}^{\infty} h(t)\, dt \int_{-\infty}^{\infty} F(s, t; \lambda)\chi_{\alpha,\beta}(s)\, ds = \int_{-\infty}^{\infty} h(t)\{G(\beta, t; \lambda) - G(\alpha, t; \lambda)\}\, dt$$

$$= \int_{-\infty}^{\infty} \chi_{\alpha,\beta}(s)\, ds \frac{d}{ds} \int_{-\infty}^{\infty} h(t) G(s, t; \lambda)\, dt$$

$$= (F_\lambda h, \chi_{\alpha,\beta}) = (h, F_\lambda \chi_{\alpha,\beta}).$$

Therefore, for almost all t,

$$(F_\lambda \chi_{\alpha,\beta})(t) = \int_{-\infty}^{\infty} \overline{F(s, t; \lambda)}\chi_{\alpha,\beta}(s)\, ds.$$

Given an arbitrary function $f \in L^2$, a sequence $\{f_n\}_1^\infty$ of step-functions with compact support can be found which converges strongly to f. The sequence $\{F_\lambda f_n\}_1^\infty$ will then converge strongly to the function $F_\lambda f$, and a sequence of subscripts $\{n_i\}_{i=1}^\infty$ can then be found such that, for almost all t,

$$\lim_{i \to \infty} (F_\lambda f_{n_i})(t) = (F_\lambda f)(t). \tag{9}$$

On the other hand, a set of measure zero can be found such that, at each point t not in this set, all the equalities

$$(F_\lambda f_{n_i})(t) = \int_{-\infty}^{\infty} \overline{F(s, t; \lambda)} f_{n_i}(s)\, ds \qquad (i = 1, 2, 3, \ldots)$$

hold simultaneously. At each of these points t the right-hand side tends to the limit

$$\int_{-\infty}^{\infty} \overline{F(s, t; \lambda)} f(s)\, ds,$$

and by (9) this limit is equal to $(F_\lambda f)(t)$. Hence the assertion (2) is proved.

3°. We shall now prove the relation (3). To do this, we take the following inequality, which is contained in (6),

$$\int_{-\infty}^{\infty} ds \left| \frac{1}{h} \int_t^{t+h} F(x, s; \lambda) k(x)\, dx \right|^2 = (F_\lambda f_{t,h}, F_\lambda f_{t,h}) \leqslant \frac{1}{\lambda^2},$$

from which it follows that

$$\int_{e} dt \int_{-\infty}^{\infty} ds \frac{1}{k^2(t)} \left| \frac{1}{h} \int_{t}^{t+h} F(x, s; \lambda) k(x)\, dx \right|^2 \leqslant \frac{1}{\lambda^2} \int_{e} \frac{dt}{k^2(t)},$$

where e is an arbitrary sub-set of finite measure of the set e(a) of all those t for which

$$1/k(t) \leqslant a.$$

Since for each s the function

$$\frac{1}{k^2} \left| \frac{1}{h} \int_{t}^{t+h} F(x, s; \lambda) k(x)\, dx \right|^2$$

tends, for almost all $t \in$ e, to the limit $|F(t, s; \lambda)|^2$ when $h \to 0$, it follows, by Fatou's theorem, that

$$\int_{e} dt \int_{-\infty}^{\infty} |F(t, s; \lambda)|^2\, ds \leqslant \frac{1}{\lambda^2} \int_{e} \frac{dt}{k^2(t)}.$$

Since the number a and the set e $\subset$ e(a) are arbitrary, we have that, for almost all t,

$$\int_{-\infty}^{\infty} |F(t, s; \lambda)|^2\, ds \leqslant \frac{1}{\lambda^2} \frac{1}{k^2(t)},$$

from which the inequality (3) follows.

The relation (3) and the formula (8), which gives a representation of the operator F_λ, do not prevent us from changing the function $F(t, s; \lambda)$, for each of almost all t, on a set of measure zero on the s-axis. We make use of this fact in order, in accordance with Lemma 2 of §25, to make the function $F(t, s; \lambda)$ measurable in the s, t-plane.

4°. It remains to prove that

$$\overline{F(s, t; \lambda)} = F(t, s; \lambda)$$

almost everywhere in the s, t-plane.

To do this, we choose functions $f \in L^2$, $g \in [L^2]_K$, and for them we write the equality

$$(F_\lambda g, f) = (g, F_\lambda f)$$

in the form

$$\int_{-\infty}^{\infty} \overline{f(s)}\, ds \int_{-\infty}^{\infty} \overline{F(t, s; \lambda)} g(t)\, dt = \int_{-\infty}^{\infty} g(t)\, dt \int_{-\infty}^{\infty} F(s, t; \lambda) \overline{f(s)}\, ds.$$

Since, by (3),

$$\int_{-\infty}^{\infty} dt \int_{-\infty}^{\infty} |g(t)| \,.\, |f(s)| \,.\, |F(t, s; \lambda)| \, ds$$

$$\leqslant \int_{-\infty}^{\infty} |g(t)| \,.\, \|f\| \frac{1}{|\lambda|} K_1(t) \, dt < \infty,$$

it follows, by Fubini's theorem, that the order of integration can be changed on the left-hand side of the last equality. Therefore

$$\int_{-\infty}^{\infty} g(t) \, dt \int_{-\infty}^{\infty} \overline{F(t, s; \lambda) f(s)} \, ds = \int_{-\infty}^{\infty} g(t) \, dt \int_{-\infty}^{\infty} F(s, t; \lambda) \overline{f(s)} \, ds,$$

which implies

$$\overline{F(s, t; \lambda)} = F(t, s; \lambda),$$

since the function $f \in L^2$ is arbitrary, and the set of functions g is dense in L^2.

This concludes the proof of Theorem 1.

We now consider some further properties of the operator F_λ.

Lemma. *Let*

$$\Omega(\lambda) = (F_\lambda f, g) \qquad (f, g \in L^2), \tag{10}$$

and let Δ *be an arbitrary interval on the* λ*-axis at a distance* $\delta > 0$ *from the point* $\lambda = 0$. *Then*

$$\operatorname{Var}_\Delta \Omega(\lambda) \leqslant \|f\| \,.\, \|g\|. \tag{11}$$

Further, if $g \in [L]_K$, *then*

$$\operatorname{Var}_\Delta \Omega(\lambda) \leqslant \frac{1}{\delta} \|f\| \,.\, \|g\|_K, \tag{12}$$

and if $f, g \in [L^2]_K$, *then*

$$\operatorname{Var}_\Delta \Omega(\lambda) \leqslant \frac{1}{\delta^2} \|f\|_K \,.\, \|g\|_K. \tag{13}$$

Proof. We choose points $\lambda_0 < \lambda_1 < \lambda_2 < \cdots < \lambda_n$ in the interval Δ, and form the expression

$$V_n \equiv \sum_{k=1}^{n} |\Omega(\lambda_k) - \Omega(\lambda_{k-1})| = \sum_{k=1}^{n} |((F_{\lambda_k} - F_{\lambda_{k-1}}) f, g)|.$$

Since $F_{\lambda_k} - F_{\lambda_{k-1}}$ is a positive operator, we have, by the Cauchy–Bunyakovski inequality

$$V_n \leqslant \sum_{k=1}^{n} \sqrt{\{((F_{\lambda_k} - F_{\lambda_{k-1}})f, f)\}} \,.\, \sqrt{\{((F_{\lambda_k} - F_{\lambda_{k-1}})g, g)\}}$$

$$\leqslant \sqrt{\left\{\sum_{k=1}^{n} ((F_{\lambda_k} - F_{\lambda_{k-1}})f, f)\right\}} \,.\, \sqrt{\left\{\sum_{k=1}^{n} ((F_{\lambda_k} - F_{\lambda_{k-1}})g, g)\right\}}$$

$$\leqslant \sqrt{\{((F_{\lambda_n} - F_{\lambda_0})f, f)\}} \,.\, \sqrt{\{((F_{\lambda_n} - F_{\lambda_0})g, g)\}} \leqslant \sqrt{\{(F_\delta f, f)(F_\delta g, g)\}}.$$

Since $\|F_\delta\| = 1$, the estimate (11) follows from the above inequality. By applying the estimate (2) to $(F_\delta g, g)$ when $g \in [L^2]_K$, we obtain (12), and finally, if $f, g \in [L^2]_K$, then, using (2) again, we obtain (13). Q.E.D.

Theorem 2. *As the kernel of the operator F_λ a function $F^\circ(s, t; \lambda)$ can be selected (from the set of functions which are equivalent with respect to (s, t) for each λ) so that, except for a set (independent of λ) of measure zero in the s, t-plane, this function is of bounded variation with respect to λ in any finite or infinite interval Δ of the λ-axis which is at a positive distance from the point $\lambda = 0$. If this distance is denoted by δ, then*

$$\mathrm{Var}_\Delta\, F^\circ(s, t; \lambda) \leqslant \frac{1}{\delta^2} K(s)K(t),$$

and if $\lambda \neq 0$, then

$$|F^\circ(s, t; \lambda)| \leqslant \frac{1}{\lambda^2} K(s)K(t).$$

Finally, for any non-zero λ, almost everywhere in the s, t-plane

$$F^\circ(s, t; \lambda - 0) = F^\circ(s, t; \lambda).$$

Proof. Let $F(s, t; \lambda)$ be some kernel of the operator F_λ; we shall construct from it a kernel $F^\circ(s, t; \lambda)$ which is equivalent to it in the s, t-plane for every λ and which satisfies the conditions of the theorem.

To do this, we introduce the function

$$\varphi_n(s; u) = \begin{cases} n/2 & \text{if } |s-u| \leqslant 1/n \\ 0 & \text{if } |s-u| > 1/n \end{cases} \qquad n = 1, 2, 3, \ldots,$$

and let

$$\varphi_n^{(a)}(s; u) = \begin{cases} \varphi_n(s; u) & \text{if } u \in \mathrm{e}(a), \\ 0 & \text{if } u \notin \mathrm{e}(a), \end{cases}$$

where $\mathrm{e}(a)$ is the set of those values of u for which $K(u) \leqslant a$. Clearly, for a fixed s, the function $\varphi_n^{(a)}(s; u) \in [L^2]_K$. Further, we put

$$\Phi_n^{(a)}(s, t; \lambda) = \iint_{-\infty}^{\infty} F(u, v; \lambda)\varphi_n^{(a)}(s; u)\varphi_n^{(a)}(t; v)\, du\, dv. \tag{14}$$

We observe that, thanks to the estimate (3), this integral exists as a double integral in the sense of absolute convergence, and not merely as a repeated integral.

By (13)

$$\operatorname{Var}_\Delta \Phi_n^{(a)}(s, t; \lambda) \leqslant \frac{1}{\delta^2} \iint_{-\infty}^{\infty} K(u)K(v)\varphi_n^{(a)}(s; u)\varphi_n^{(a)}(t; v)\, du\, dv \leqslant \frac{a^2}{\delta^2}. \tag{15}$$

Thus all the functions $\Phi_n^{(a)}(s, t; \lambda)$ $(n = 1, 2, 3, \ldots)$ are of uniformly bounded variation with respect to λ in the interval Δ. Consequently, by using Helly's theorem, in combination with the repeated diagonal-process, we find a sequence $\{n_k\}_{k=1}^{\infty}$ depending on an arbitrarily chosen point (s, t) and such that, for all $\lambda \neq 0$, the limit

$$\lim_{k \to \infty} \Phi_{n_k}^{(a)}(s, t; \lambda) \equiv \Phi^{(a)}(s, t; \lambda)$$

exists. Moreover, by (15), we have, for any s, t

$$\operatorname{Var}_\Delta \Phi^{(a)}(s, t; \lambda) \leqslant \frac{a^2}{\delta^2}. \tag{16}$$

This last inequality can be sharpened, as follows. Let $e_2^*(a)$ be the set of all those points $(s, t) \in e(a) \times e(a) \equiv e_2(a)$ at which

$$\lim_{n \to \infty} \iint_{-\infty}^{\infty} K(u)K(v)\varphi_n^{(a)}(s; u)\varphi_n^{(a)}(t; v)\, du\, dv = K(s)K(t).$$

By known theorems, $\operatorname{meas}\{e_2(a) \setminus e_2^*(a)\} = 0$.

When $(s, t) \in e_2^*(a)$, we can take the limit as $n \to \infty$ of the middle member of the inequality (15); and it then follows that

$$\operatorname{Var}_\Delta \Phi^{(a)}(s, t; \lambda) \leqslant \frac{1}{\delta^2} K(s)K(t) \qquad (s, t \in e_2^*(a)). \tag{17}$$

This is the sharpened version of (16).

On the other hand, by (14), we have, for every non-zero λ, almost everywhere in $e_2(a)$,

$$\lim_{n \to \infty} \Phi_n^{(a)}(s, t; \lambda) = F(s, t; \lambda),$$

and therefore, for every non-zero λ and almost everywhere in $e_2(a)$,

$$\Phi^{(a)}(s, t; \lambda) = F(s, t; \lambda). \tag{18}$$

We now express the s, t-plane E_2 as the sum of disjoint sets:

$$E_2 = \bigcup_{a=1}^{\infty} E_2(a),$$

where

$$E_2(1) = e_2(1), \qquad E_2(a+1) = e_2(a+1) \setminus e_2(a),$$

and we define a function $F^{\circ}(s, t; \lambda)$ by putting

$$F^{\circ}(s, t; \lambda) = \Phi^{(a)}(s, t; \lambda) \qquad \text{when} \qquad (s, t) \in E_2(a).$$

It is clear that, by (18), for every λ and almost everywhere in the s, t-plane

$$F^{\circ}(s, t; \lambda) = F(s, t; \lambda),$$

and therefore $F^{\circ}(s, t; \lambda)$ is a kernel of the operator F_λ, *i.e.*, for any $f, g \in L^2$,

$$\iint_{-\infty}^{\infty} F^{\circ}(s, t; \lambda) f(t) \overline{g(s)}\, ds\, dt = (F_\lambda f, g).$$

Also, on the set

$$E_2^* = \bigcup_{a=1}^{\infty} (E_2(a) \cap e_2^*(a)),$$

i.e., on the whole s, t-plane except for a set of measure zero which does not depend on λ, because of (17) the estimate

$$\operatorname{Var}_\Delta F^{\circ}(s, t; \lambda) \leq \frac{1}{\delta^2} K(s) K(t)$$

holds. But this implies, in particular, that for non-zero λ and almost everywhere in the s, t-plane

$$|F^{\circ}(s, t; \lambda)| \leq \frac{1}{\lambda^2} K(s) K(t).$$

We now prove that, for every non-zero λ and almost everywhere in the s, t-plane

$$F^{\circ}(s, t; \lambda - 0) = F^{\circ}(s, t; \lambda). \tag{19}$$

To do this, it suffices to observe, firstly, that for $(s, t) \in E_2^*$ the limit

$F^{\circ}(s, t; \lambda-0)$ exists for any λ, and secondly, that for any $f, g \in [L^2]_K$,

$$|F^{\circ}(s, t; \mu)f(t)\overline{g(t)}| \leqslant \frac{1}{\mu^2} K(s)K(t)\,|f(s)|\,.\,|g(t)|,$$

and here the right-hand side is integrable in the whole s, t-plane, and therefore, when $\lambda \neq 0$, we can pass to the limit under the integral sign in the following equality:

$$\iint_{-\infty}^{\infty} F^{\circ}(s, t; \lambda)f(t)\overline{g(s)}\, ds\, dt = (F_\lambda f, g) = (F_{\lambda-0} f, g)$$

$$= \lim_{\mu \to \lambda - 0} \iint_{-\infty}^{\infty} F^{\circ}(s, t; \mu)f(t)\overline{g(s)}\, ds\, dt$$

$$= \iint_{-\infty}^{\infty} F^{\circ}(s, t; \lambda-0)f(t)\overline{g(s)}\, ds\, dt.$$

From this the relation (19) follows, and the theorem is completely proved.

From now on we shall take the kernel $F(s, t; \lambda)$ of the operator F_λ to be this kernel $F^{\circ}(s, t; \lambda)$.

Remark. By a similar argument, using the estimate (12) instead of (13), it is easily proved that, for any function $f \in L^2$,

$$\mathrm{Var}_\Delta \int_{-\infty}^{\infty} F(s, t; \lambda)f(t)\, dt \leqslant \frac{1}{\delta} K(s)\,.\,\|f\|,$$

except for a set of measure zero, not depending on Δ, on the s-axis.

§119. Spectral representation of a Carleman kernel

Theorem 1. Let $\Delta = [\alpha, \beta)$ be a finite interval situated at a positive distance from the point $\lambda = 0$, and let

$$F(s, t; \Delta) = F(s, t; \beta) - F(s, t; \alpha).$$

Then, almost everywhere in the s, t-plane

$$\int_{-\infty}^{\infty} K(s, u)F(u, t; \Delta)\, du = \int_{\alpha}^{\beta} \lambda\, d_\lambda F(s, t; \lambda). \tag{1}$$

(The excluded set of measure zero in the s, t-plane on which (1) *does not hold depends, in general, on* α *and* β.)

The theorem means that (1) holds for every $s \notin e_{\alpha,\beta}$ and for almost all t ($e_{\alpha,\beta}$ is some set of measure zero on the s-axis).

Proof. For any $f, g \in L^2$ we have

$$(A^*E(\Delta)f, g) = \int_\alpha^\beta \lambda \, d_\lambda (E_\lambda f, g),$$

which, clearly, may be rewritten as

$$\int_{-\infty}^{\infty} \overline{g(s)} \, ds \int_{-\infty}^{\infty} K(s, u) \left\{ \int_{-\infty}^{\infty} F(u, t; \Delta) f(t) \, dt \right\} du$$

$$= \int_\alpha^\beta \lambda \, d \left\{ \int_{-\infty}^{\infty} \overline{g(s)} \, ds \int_{-\infty}^{\infty} F(s, t; \lambda) f(t) \, dt \right\}.$$

Let us assume that $f, g \in [L^2]_{K_1}$. Then by Theorem 2 of §118, the right-hand side is equal to

$$\iint_{-\infty}^{\infty} f(t)\overline{g(s)} \left\{ \int_\alpha^\beta \lambda d_\lambda F(s, t; \lambda) \right\} dt \, ds$$

But, by Theorem 1 of §118, almost everywhere

$$\int_{-\infty}^{\infty} |K(s, u) F(u, t; \Delta)| \, du \leq 2K(s) \frac{K_1(t)}{\delta},$$

where δ is the distance of the interval Δ from the point $\lambda = 0$. Therefore the order of integration can be changed on the left-hand side, and so

$$\iint_{-\infty}^{\infty} f(t)\overline{g(s)} \left\{ \int_{-\infty}^{\infty} K(s, u) F(u, t; \Delta) \, du \right\} dt \, ds$$

$$= \iint_{-\infty}^{\infty} f(t)\overline{g(s)} \left\{ \int_\alpha^\beta \lambda d_\lambda F(s, t; \lambda) \right\} dt \, ds$$

for any $f, g \in [L^2]_{K_1}$. Since $[L^2]_{K_1}$ is dense in L^2, it follows from this that

$$\int_{-\infty}^{\infty} K(s, u) F(u, t; \Delta) \, du = \int_\alpha^\beta \lambda d_\lambda F(s, t; \lambda)$$

almost everywhere in the s, t-plane. Q.E.D.

Theorem 2. *For almost all s, the formula*

$$\int_{-\infty}^{\infty} K(s, u)F(u, t; \Delta)\, du = \int_{\alpha}^{\beta} \lambda d_\lambda F(s, t; \lambda)$$

is also valid when $\beta = \infty$ and $\alpha > 0$ (or when $\alpha = -\infty$ and $\beta < 0$) if the right-hand side is regarded as the l.i.m. *with respect to t when $\beta \to \infty$ (or when $\alpha \to -\infty$) through an arbitrarily chosen sequence.*

Proof. Assuming that $\alpha > 0$ and that β is finite, we re-write formula (1) in the form

$$\int_{\alpha}^{\beta} \lambda d_\lambda \overline{F(s, t; \lambda)} = \int_{-\infty}^{\infty} \overline{K(s, u)}\,\overline{F(u, t; \Delta)}\, du$$

$$= \int_{-\infty}^{\infty} F(t, u; \Delta)K(u, s)\, du = (F_\beta - F_\alpha)K_s, \tag{2}$$

where the function $K_s(t) = K(t, s)$ belongs to L^2 for almost all s.

We choose a sequence of numbers β_k tending to infinity. Then, by Theorem 1, we can find a set e_α of measure zero and depending only on α, and such that, for any $s \notin e_\alpha$, all the equalities (2) with $\beta = \beta_k$ $(k = 1, 2, 3, \ldots)$ will hold simultaneously almost everywhere on the t-axis. Hence it follows that, for $s \notin e_\alpha$ and $k = 1, 2, 3, \ldots,$

$$\left\| \int_{\alpha}^{\beta_k} \lambda d_\lambda \overline{F(s, t; \lambda)} + F_\alpha K_s \right\| = \left\| E_{\beta_k} K_s - K_s \right\|.$$

The right-hand side tends to 0 as $k \to \infty$. Thus, for all $s \notin e_\alpha$,

$$-\int_{-\infty}^{\infty} F(t, u; \alpha)K(u, s)\, du = \underset{k \to \infty}{\text{l.i.m.}} \int_{\alpha}^{\beta_k} \lambda d_\lambda \overline{F(s, t; \lambda)},$$

and therefore, for almost all s,

$$-\int_{-\infty}^{\infty} K(s, u)F(u, t; \alpha)\, du = \underset{k \to \infty}{\text{l.i.m.}} \int_{\alpha}^{\beta_k} \lambda d_\lambda F(s, t; \lambda),$$

from which the theorem follows.

Theorem 3. *For almost all s,*

$$K(t, s) = \text{l.i.m.} \int_{-\infty}^{\infty} \lambda d_\lambda F(t, s; \lambda),$$

where the right-hand side is the l.i.m. *with respect to t of an improper integral with singular points at* $\lambda=-\infty, 0, \infty$; *in other words, for almost all* s,

$$\lim \int_{-\infty}^{\infty} \left| K(t, s) - \left(\int_{-N'}^{-\varepsilon'} + \int_{\varepsilon}^{N} \right) \lambda d_\lambda F(t, s; \lambda) \right|^2 dt = 0,$$

where $\varepsilon, \varepsilon' \to 0$ *and* $N, N' \to \infty$ *through arbitrarily chosen sequences of positive numbers.*

In the same sense and for almost all s

$$K(s, t) = \text{l.i.m.} \int_{-\infty}^{\infty} \lambda d_\lambda F(s, t; \lambda).$$

Proof. We choose arbitrarily sequences of positive numbers $\varepsilon_i, \varepsilon_i' \to 0$ $(\varepsilon_i, \varepsilon_i' < 1)$ and $N_k, N_k' \to \infty$ $(N_k, N_k' > 1)$.

Then, as in the proof of Theorem 2, there is a set e of measure zero such that, for every $s \notin \mathrm{e}$, the following equalities will hold simultaneously for all i, k, l, m and almost everywhere on the t-axis:

$$\left(\int_{-N_k'}^{-\varepsilon_i'} + \int_{\varepsilon_l}^{N_m} \right) \lambda d_\lambda F(t, s; \lambda) = (F_{-\varepsilon_i'} - F_{-N_k'}) K_s + (F_{N_m} - F_{\varepsilon_l}) K_s,$$

or

$$\left(\int_{-N_k'}^{-\varepsilon_i'} + \int_{\varepsilon_l}^{N_m} \right) \lambda d_\lambda F(t, s; \lambda) - E_0 K_s + E_{+0} K_s - K_s$$
$$= (E_{-\varepsilon_i'} - E_0) K_s + (E_{+0} - E_{\varepsilon_l}) K_s + (F_{M_m} - F_{-N_k'}) K_s.$$

For every $s \notin \mathrm{e}$ the right-hand side, as a function of t, tends to zero in the L^2-metric when $i, k, l, m \to \infty$. Consequently, for $s \notin \mathrm{e}$,

$$K_s(t) - (E_{+0} - E_0) K_s = \text{l.i.m.} \left(\int_{-N_k'}^{-\varepsilon_i'} + \int_{\varepsilon_l}^{N_m} \right) \lambda d_\lambda F(t, s; \lambda).$$

It remains to verify that

$$\|(E_{+0} - E_0) K_s\| = 0 \qquad \text{for almost all } s.$$

But, for any $f \in L^2$,

$$(f, (E_{+0} - E_0) K_s) = ((E_{+0} - E_0) f, K_s)$$
$$= \int_{-\infty}^{\infty} \overline{K_s(t)} ((E_{+0} - E_0) f)(t)\, dt = A^*(E_{+0} - E_0) f = 0.$$

§120. Generalization of the Hilbert–Schmidt formula

Theorem 1. *For any function $f \in L^2$ and for almost all s*

$$\int_{-\infty}^{\infty} K(s, t)f(t)\,dt = \int_{-\infty}^{\infty} \lambda d_\lambda \left\{\int_{-\infty}^{\infty} F(s, t; \lambda)f(t)\,dt\right\},$$

the integral with respect to λ being improper as regards the points $\lambda = -\infty, 0, \infty$.

Proof. By Theorem 3 of §119, for any $f \in L^2$, the following equality holds for almost all s:

$$\int_{-\infty}^{\infty} K(s, t)f(t)\,dt = \int_{-\infty}^{\infty} f(t)\,dt \int_{-\infty}^{\infty} \lambda d_\lambda F(s, t; \lambda)$$

$$\equiv \lim_{\substack{\varepsilon_i', \varepsilon_l \to 0 \\ N_k', N_m \to \infty}} \int_{-\infty}^{\infty} f(t)\,dt \left(\int_{-N_k'}^{-\varepsilon_i'} + \int_{\varepsilon_l}^{N_m}\right) \lambda d_\lambda F(s, t; \lambda).$$

We have only to prove that the order of integration can be changed on the right-hand side of this equality. To do this, we observe that

$$\mathrm{Var}_\Delta \int_{-\infty}^{\infty} F(s, t; \lambda)f(t)\,dt \leq \frac{1}{\delta} K(s) \,.\, \|f\|$$

by the remark at the end of §118, and therefore, for almost all s, when $\alpha\beta > 0 \quad (\beta > \alpha)$ the Stieltjes integral

$$\int_{\alpha}^{\beta} \lambda d_\lambda \int_{-\infty}^{\infty} F(s, t; \lambda)f(t)\,dt \tag{1}$$

exists. Also, for almost all s, the integral

$$\int_{-\infty}^{\infty} f(t)\,dt \int_{\alpha}^{\beta} \lambda d_\lambda F(s, t; \lambda) \tag{2}$$

exists for any $f(t) \in L^2$, since, by virtue of the formula (2) in §119,

$$\int_{\alpha}^{\beta} \lambda d_\lambda F(s, t; \lambda) \in L_t^2(-\infty, \infty).$$

If $f(t) \in [L^2]_K$, then the equality of these integrals follows from an estimate of the quantity

$$J_n = \int_{-\infty}^{\infty} f(t)\,dt \left\{\int_{\alpha}^{\beta} \lambda d_\lambda F(s, t; \lambda) - \sum_{k=1}^{n} \lambda_k F(s, t; \Delta_k)\right\},$$

where, in the braces, there is the difference between the Stieltjes integral and its integral sums. For, by Theorem 2 of §118,†

$$|J_n| \leqslant \max_{1\leqslant k\leqslant n} |\Delta_k| \int_{-\infty}^{\infty} |f(t)| \, [\mathrm{Var}_{[\alpha,\beta)} F(s, t; \lambda)] \, dt$$

$$\leqslant \max_{1\leqslant k\leqslant n} |\Delta_k| \frac{1}{\delta^2} K(s) \, \|f\|_K,$$

and therefore

$$\lim_{n\to\infty} J_n = 0,$$

from which the equality of the integrals (1) and (2) follows.

For an arbitrary function $f(t) \in L^2$ a sequence of functions

$$f_n(t) \in [L^2]_K \qquad (n = 1, 2, 3, \ldots)$$

can be chosen which converges to f in L^2, and then the excluded set e of measure zero on the s-axis can be found, common for all these functions $f_n(t)$. But then by Helly's theorem on passage to the limit in a Stieltjes integral, we have, for $s \notin e$,

$$\int_{\alpha}^{\beta} \lambda d_{\lambda} \int_{-\infty}^{\infty} F(s, t; \lambda) f_n(t) \, dt \to \int_{\alpha}^{\beta} \lambda d_{\lambda} \int_{-\infty}^{\infty} F(s, t; \lambda) f(t) \, dt.$$

Moreover

$$\int_{-\infty}^{\infty} f_n(t) \, dt \int_{\alpha}^{\beta} \lambda d_{\lambda} F(s, t; \lambda) \to \int_{-\infty}^{\infty} f(t) \, dt \int_{\alpha}^{\beta} \lambda d_{\lambda} F(s, t; \lambda).$$

Therefore the integrals (1) and (2) are equal for any function $f(t) \in L^2$ when $s \notin e$. Q.E.D.

§121. Characteristic properties of Carleman integral operators

Theorem.‡ *An integral operator T in L^2 with its domain dense in L^2 is an integral operator with a Carleman kernel if and only if there is a finite, measurable function $P(s) \geqslant 0$ satisfying the following conditions:*

1°. $[L^2]_P \subseteq D(T^*)$;
2°. $\|T^* g\| \leqslant \|g\|_P$ *for any* $g \in [L^2]_P$;
3°. $(T^* f, g) = (f, T^* g)$ *for any* $f, g \in [L^2]_P$.

† Thanks to the properties of the function $F(s, t; \lambda)$ which are indicated in this theorem, its variation relative to λ can be calculated using only a partition of the interval $[\alpha, \beta)$ defined by means of rational points. Therefore $\mathrm{Var}_{[\alpha,\beta)} F(s, t; \lambda)$ is measurable in the s, t-plane.

‡ See Korotkov [56]. We prove this theorem in a rather different, but equivalent, formulation.

Proof. *Necessity.* Let T be an integral operator with a Carleman kernel $K(s, t)$. We shall prove that the conditions 1°–3° are then satisfied when $P(s) = K(s)$. To do this, we introduce the operators $A_0 \subseteq A \subseteq A^*$ associated with the kernel $K(s, t)$ as in §116. Every integral operator with kernel $K(s, t)$ is a restriction of the operator A^*. So $T \subseteq A^*$. Therefore $T^* \supseteq A^{**} \supseteq A$, and so $[L^2]_K \subseteq D(T^*)$, *i.e.*, the condition 1° has been proved.

The condition 3° is proved in a similar way, because, if $f, g \in [L^2]_K$,

$$(T^*f, g) = (A_0 f, g) = (f, A_0 g) = (f, T^* g).$$

Finally, the condition 2° follows from the fact that, for any $f, g \in [L^2]_K$,

$$\begin{aligned}(T^*f, T^*g) &= (A_0 f, A_0 g)\\ &= \int_{-\infty}^{\infty} ds \int_{-\infty}^{\infty} K(s, u) f(u)\, du \int_{-\infty}^{\infty} \overline{K(s, v)}\,\overline{g(v)}\, dv,\end{aligned}$$

and therefore

$$\begin{aligned}|(T^*f, T^*g)| &\leq \int_{-\infty}^{\infty} |f(u)|\, du \int_{-\infty}^{\infty} |g(v)|\, dv \int_{-\infty}^{\infty} |K(s, u) K(s, v)|\, ds\\ &\leq \int_{-\infty}^{\infty} |f(u)|\, K(u)\, du \int_{-\infty}^{\infty} |g(v)|\, K(v)\, dv = \|f\|_K\, \|g\|_K.\end{aligned}$$

Sufficiency. Suppose that the function $P(s)$ required by the theorem does exist. We select some element $f \in D(T)$ and an arbitrary element $g \in [L^2]_P$. The scalar product $(g, Tf) = (T^*g, f)$ is, by 1° and 2° a linear functional of g in $[L^2]_P \subseteq L_P$, with a norm $\leq \|f\|$. This functional can be extended by continuity on to the whole of the space L_P. This means that there is a measurable function $\alpha(s)$ such that

$$(g, Tf) = \int_{-\infty}^{\infty} g(s) \overline{\alpha(s)}\, ds$$

and

$$\operatorname{ess\,max} \frac{|\alpha(s)|}{P(s)} \leq \|f\|.$$

Since

$$(Tf)(s) = \alpha(s) \qquad \text{almost everywhere},$$

we have that, for almost all s,

$$|(Tf)(s)| \leq P(s)\, \|f\|. \tag{1}$$

Thus this inequality holds for the chosen function $f \in D(T)$ for any $s \notin e_f$, where e_f is a set of measure zero depending on f. We now select a countable set, everywhere dense in $D(T)$, and by orthonormalizing it we obtain an orthonormal sequence $\{h_k\}_1^\infty \subset D(T)$, and then we choose a set N of elements of the form

$$f_n = c_1 h_1 + c_2 h_2 + \cdots + c_n h_n,$$

where n runs through the set of natural numbers, and for each n the coefficients $c_1, c_2, \ldots, c_n$ take all possible complex values which have rational components.

This set N is countable. Each element $f_n \in N$ has its own corresponding excluded set e_{f_n} of measure zero on the s-axis. Let e denote the union of all these sets e_{f_n}; it too is of measure zero. Thus, if $s \notin e$, then, for any $f \in N$,

$$|(Tf)(s)| \leqslant P(s)\,\|f\|.$$

This inequality will hold, when $s \notin e$, for any element f^* belonging to the linear envelope M of the set $\{h_k\}_1^\infty$. For, let

$$f^* = \gamma_1 h_1 + \gamma_2 h_2 + \cdots + \gamma_n h_n \in \mathrm{M}.$$

We choose an element

$$f_n = c_1 h_1 + c_2 h_2 + \cdots + c_n h_n \in N.$$

Then

$$\begin{aligned} |(Tf^*)(s) - (Tf_n)(s)| &\leqslant n \,.\, \max_{1 \leqslant k \leqslant n} |\gamma_k - c_k| \,.\, \max_{1 \leqslant i \leqslant n} |(Th_i)(s)| \\ &\leqslant n \,.\, \max_{1 \leqslant k \leqslant n} |\gamma_k - c_k| \,.\, P(s), \end{aligned}$$

and this implies that, with a suitable choice of the numbers $c_1, c_2, \ldots, c_n$, we shall have the inequality

$$|(Tf^*)(s) - (Tf_n)(s)| \leqslant \varepsilon P(s) \qquad (s \notin \mathrm{e}).$$

Hence it follows that

$$\begin{aligned} |(Tf^*)(s)| &\leqslant |(Tf_n)(s)| + \varepsilon P(s) \leqslant P(s)\,\|f_n\| + \varepsilon P(s) \\ &\leqslant P(s)\,\|f^*\| + 2\varepsilon P(s), \end{aligned}$$

and since $\varepsilon > 0$ is arbitrary, our assertion is proved, *i.e.*, the inequality (1) holds when $s \notin \mathrm{e}$ for any element $f \in \mathrm{M}$. But in that case $(Tf)(s)$ is, for any $s \notin \mathrm{e}$, a linear functional in $\mathrm{M} \subset \mathrm{L}^2$ with norm $\leqslant P(s)$. This linear functional can be extended on to the whole space L^2 without increasing the norm. This implies that there is a function $K_s(t) \equiv K(s, t)$, measurable with respect to t for each $s \notin e$, and such that, for any $f \in M$ and any $s \notin e$,

$$(Tf)(s) = \int_{-\infty}^{\infty} K(s, t)f(t)\,dt, \tag{2}$$

and

$$\left\{\int_{-\infty}^{\infty} |K(s, t)|^2\,dt\right\}^{1/2} \leq P(s). \tag{3}$$

It follows from (3) that the integral

$$\int_{-\infty}^{\infty} K(s, t)h(t)\,dt \qquad (s \notin e)$$

exists, for any function $h(t) \in L^2$. But since every function $h(t) \in L^2$ is the limit of a sequence of functions $f(t) \in M$ for which the integral under consideration is a measurable function of s, so this integral is a measurable function of s for any function $h(t) \in L^2$. Hence it follows, by Lemma 2 of §25 that the function $K(s, t)$ can be regarded as measurable in the s, t-plane.

We prove that T is an integral operator having the kernel $K(s, t)$. Let $f \in M$, and $g \in [L^2]_P$. By (2)

$$(f, T^*g) = (Tf, g) = \int_{-\infty}^{\infty} \overline{g(s)}\,ds \int_{-\infty}^{\infty} K(s, t)f(t)\,dt. \tag{4}$$

On the other hand, for any element $f \in L^2$ and with any $g \in [L^2]_P$, we have, by (3),

$$\int_{-\infty}^{\infty} |g(s)|\,ds \int_{-\infty}^{\infty} |K(s, t)| \,.\, |f(t)|\,dt \leq \|f\| \int_{-\infty}^{\infty} P(s)\,|g(s)|\,ds < \infty,$$

and therefore by Fubini's theorem

$$\int_{-\infty}^{\infty} \overline{g(s)}\,ds \int_{-\infty}^{\infty} K(s, t)f(t)\,dt = \int_{-\infty}^{\infty} f(t)\,dt \int_{-\infty}^{\infty} K(s, t)\overline{g(s)}\,ds. \tag{5}$$

If $f \in M$, then, by (4), the left-hand side, and so also the right-hand side, of (5) are equal to (f, T^*g). Hence we conclude, since M is dense in L^2, that, almost everywhere,

$$(T^*g)(t) = \int_{-\infty}^{\infty} \overline{K(s, t)}g(s)\,ds \qquad (g \in [L^2]_P).$$

Therefore the right-hand side of (5) is equal to (f, T^*g) for any $f \in L^2$, and so the equality (4), and with it the representation (2) as well, are valid for any $f \in D(T)$. We have thus proved that T is an integral operator with $K(s, t)$ as its kernel.

It remains to show that, almost everywhere in the s, t-plane,

$$K(s, t) = \overline{K(t, s)}. \tag{6}$$

To do this, we use condition (3) of the theorem and the equalities (4) and (5) for arbitrary elements $f, g \in [L^2]_P$. We obtain the relation

$$\int_{-\infty}^{\infty}\int_{-\infty}^{\infty} K(s, t)\overline{g(s)}f(t)\,ds\,dt = \int_{-\infty}^{\infty} \overline{g(s)}\,ds \int_{-\infty}^{\infty} K(s, t)f(t)\,dt$$

$$= (f, T^*g) = (T^*f, g) = \overline{(g, T^*f)}$$

$$= \overline{\int_{-\infty}^{\infty} \overline{f(t)}\,dt \int_{-\infty}^{\infty} K(t, s)g(s)\,ds} = \int_{-\infty}^{\infty}\int_{-\infty}^{\infty} \overline{K(t, s)}\overline{g(s)}f(t)\,ds\,dt,$$

from which it follows that the equality (6) holds almost everywhere in the s, t-plane. Q.E.D.

Remark. (See Akhiezer [3].) The conditions 1°–3° of the theorem just proved, together with the requirement

4°. $|(T^*f, g)| \leq \|f\|_P \, \|g\|_P \qquad (f, g \in [L^2]_P)$

are necessary and sufficient conditions for the operator T to be an integral operator with a Carleman kernel $K(s, t)$ which satisfies the inequality

$$|K(s, t)| \leq P(s)P(t)$$

almost everywhere in the s, t-plane. Examples of such operators are the operators F_λ $(\lambda \neq 0)$ introduced in §118 for an arbitrary Carleman operator.

§122. Von Neumann's theorem

A natural question to ask is: how wide is the class of operators in L^2 which are either Carleman integral operators or are unitarily equivalent to them? The answer was given by von Neumann [95]. His theorem reads:

a self-adjoint operator B in L^2 is unitarily equivalent to an integral operator with a Carleman kernel if and only if the continuous spectrum of the operator B contains the point $\lambda = 0$.

Proof. We begin by proving the sufficiency of the condition, and in doing so we may assume that the operator B has a pure point spectrum. For, by the theorems in §94, the operator $C = B + R$ will in any case have a pure point spectrum if a suitable self-adjoint operator R with a finite (or even an arbitrarily small) absolute norm is chosen; an operator R of this kind and all operators unitarily equivalent to it are Hilbert–Schmidt integral operators.

Thus we can assume that the operator has a pure point spectrum. Let $\{\lambda_i\}_1^\infty$ be the sequence of all its eigenvalues. Let us assume that the point $\lambda = 0$ is a limit-point of the spectrum, so that there is an infinite sequence of subscripts m_q $(\pm q = 1, 2, 3, \ldots)$ such that

$$|\lambda_{m_q}| \leqslant \frac{1}{1+|q|} \qquad (\pm q = 1, 2, 3, \ldots).$$

Let $\{n_p\}_{p=-\infty}^\infty$ be the set of the remaining subscripts. We express the set of all natural numbers in the form of a countable set of sequences

$$\{l_{pk}\}_{p=-\infty}^\infty \qquad (k = 1, 2, 3, \ldots)$$

where

$$l_{p1} = n_p, \qquad l_{pk} = m_{(2p-1)2^{k-2}} \qquad (k = 2, 3, \ldots).$$

Then

$$\sum_{k=1}^\infty |\lambda_{l_{pk}}| \leqslant |\lambda_{n_p}| + \sum_{k=2}^\infty \frac{1}{|2p-1|\, 2^{k-2}} = |\lambda_{n_p}| + \frac{2}{|2p-1|} = \mathscr{C}_p < \infty.$$

We select some complete orthonormal system of functions $\{\psi_k(s)\}_1^\infty$ in $L^2(0, 1)$ which satisfy the condition for uniform boundedness

$$|\psi_k(s)| \leqslant \mathscr{M} \qquad (0 \leqslant s \leqslant 1,\ k = 1, 2, 3, \ldots).$$

For example, we could take $\psi_k(s) = \sqrt{2} \,.\, \sin k\pi s$. Further, we put

$$\varphi_{p,k}(s) = \begin{cases} \psi_k(s-p) & s \in [p, p+1), \\ 0 & s \notin [p, p+1). \end{cases}$$

Clearly, $\{\varphi_{p,k}(s)\}$ is a complete orthonormal system in $L^2(-\infty, \infty)$. The series

$$K(s, t) = \sum_{p=-\infty}^\infty \sum_{k=1}^\infty \lambda_{l_{p,k}} \varphi_{p,k}(s) \varphi_{p,k}(t)$$

converges everywhere, and moreover converges absolutely. For, if, for example,

$$i \leqslant s < i+1 \qquad \text{and} \qquad j \leqslant t < j+1,$$

then, if $i \neq j$, all the terms are equal to 0, and if $i = j$,

$$K(s, t) = \sum_{k=1}^\infty \lambda_{l_{ik}} \psi_k(s-i) \psi_k(t-i),$$

and so

$$|K(s,t)| \leq \mathcal{M}^2 \sum_{k=1}^{\infty} |\lambda_{l_{ik}}| \leq \mathcal{M}^2 \mathcal{E}_i \qquad (i \leq s, t < i+1).$$

We also observe that

$$K^2(s) \equiv \int_{-\infty}^{\infty} |K(s,t)|^2 \, dt \leq \mathcal{M}^4 \mathcal{E}_i^2 \qquad (i \leq s < i+1).$$

Therefore this quantity is finite for any s. Thus we have arrived at the integral operator

$$(Af)(t) = \int_{-\infty}^{\infty} K(s,t) f(s) \, ds$$

with a Carleman kernel, and our construction shows that

$$(A\varphi_{p,k})(t) = \lambda_{l_{p,k}} \varphi_{p,k}(t).$$

Since $\{\varphi_{p,k}(t)\}$ is a complete orthonormal sequence, the spectrum of the constructed integral operator is a pure point spectrum, and the sequence of points $\lambda_{l_{pk}}$, where $\pm p = 0, 1, 2, \ldots$, $k = 1, 2, 3, \ldots$, constitutes the complete set of its eigenvalues.

This completes the proof of the sufficiency of the condition.

To prove its necessity, let us assume that we are given a self-adjoint integral operator

$$(Af)(t) = \int_{-\infty}^{\infty} K(s,t) f(t) \, dt$$

with a Carleman kernel.

We have to prove that the point 0 belongs to its continuous spectrum. To do this, it suffices (see §93) to construct an orthonormal system (not necessarily complete) of functions $\{\varphi_k(t)\}_1^{\infty} \subset L^2(-\infty, \infty)$ such that

$$\lim_{k \to \infty} \|A\varphi_k\| = 0.$$

The possibility of this construction, and hence the necessity of the condition in von Neumann's theorem, follows from the next lemma.

Lemma. *Given any finite system of functions $\{\varphi_k(t)\}_1^m \subset L^2(-\infty, \infty)$, and any positive number ε, there is a function $\psi(t) \in L^2$ which satisfies the conditions*

$$\|\psi\| = 1, \qquad \|A\psi\| < \varepsilon, \qquad (\varphi_k, \psi) = 0 \qquad (k = 1, 2, \ldots, m),$$

where A is an integral operator (not necessarily self-adjoint) with a Carleman kernel.

Proof. By one of the properties of Carleman kernels, we can find a number $N>0$ and a set $E\subset[0, 1]$ of positive Lebesgue measure such that the kernel $K(s, t)$ satisfies the inequality

$$\int_{-\infty}^{\infty} |K(s, t)|^2\, dt \leq N \qquad (s\in E).$$

We put

$$w_s(t)=\begin{cases}\overline{K(s, t)} & (s\in E),\\ 0 & (s\notin E),\end{cases}$$

and so

$$\|w_s\|^2\leq N \qquad (s\in E),$$
$$\|w_s\|^2=0 \qquad (s\notin E),$$

and therefore

$$\int_{-\infty}^{\infty} \|w_s\|^2\, ds=\int_E \|w_s\|^2\, ds\leq N.$$

We now introduce a new measure, equal to

$$\int_e \|w_s\|^2\, ds$$

for the set e, and we call it, for brevity, *w-measure.*

The next step in the construction consists in choosing some orthonormal basis $\{g_k(t)\}_1^\infty\subset L^2$, by means of which we construct an infinite sequence of functions

$$\Phi_k(s)=(w_s, g_k) \qquad (k=1, 2, \dots);$$

all of these differ from 0 only when $s\in E$. By Lusin's theorem, we can remove from E some set of arbitrarily small w-measure $\leq\eta$ (we shall take $\eta<\varepsilon^2/4$) so that on the rest of the set the function

$$\Phi_0(s)=\|w_s\|^2$$

will be continuous. From this last set we construct a sub-set of w-measure $<\eta/2$ so that on the new set the function $\Phi_1(s)$ will also be continuous. By continuing this process indefinitely, we arrive at a set $F\subset E$ such that the complement of F in E has w-measure $\leq 2\eta$. Clearly, the set F can be regarded as closed, because otherwise it can be closed by removing one more open set of arbitrarily small w-measure. On this set F all the functions $\Phi_k(s)$ $(k=0, 1, 2, \dots)$ are continuous. Therefore, if $s\in F$, $s_0\in F$, and if $s\to s_0$, then $\|w_s\|\to\|w_{s_0}\|$ and $(w_s, g_k)\to(w_{s_0}, g_k)$ $(k=1, 2, 3, \dots)$. It follows from these relations that

$$\lim_{s\to s_0}\|w_s-w_{s_0}\|=0.$$

Moreover, because the set F is bounded and closed, this holds uniformly; that is, for any $\theta>0$ we can find a $\delta>0$ such that, when $|s-s_0|<\delta$, and $s'', s' \in \mathrm{F}$, the inequality

$$\|w_{s''}-w_{s'}\|^2<\theta$$

holds. Therefore, for any $\theta>0$ (we shall take $\theta<\varepsilon^2/2$), we can find points $s_1, s_2, \ldots, s_p \in \mathrm{F}$, where $p=p(\theta)$, such that

$$\|w_s-w_{s_j}\|^2<\theta$$

for any point $s\in\mathrm{F}$ and for any j $(1\leqslant j\leqslant p)$.

We now choose an arbitrary function $\psi(t)\in \mathrm{L}^2$ such that $\|\psi\|=1$ and such that ψ is orthogonal to the functions $w_{s_k}(t)$ $(k=1, 2, \ldots, p)$, and we shall show that

$$\int_{-\infty}^{\infty} |(w_s, \psi)|^2\, ds<\varepsilon^2.$$

For, we have

$$\int_{-\infty}^{\infty} |(w_s, \psi)|^2\, ds=\int_{\mathrm{E}} |(w_s, \psi)|^2\, ds$$

$$=\int_{\mathrm{F}} |(w_s, \psi)|^2\, ds+\int_{\mathrm{E-F}} |(w_s, \psi)|^2\, ds$$

$$=\int_{\mathrm{F}} |(w_s-w_{s_j}, \psi)|^2\, ds+\int_{\mathrm{E-F}} |(w_s, \psi)|^2\, ds\leqslant\theta+2\eta<\varepsilon^2.$$

Let us assume additionally that $\psi(s)=0$ for $s\notin \mathrm{E}$. Then

$$\|\mathrm{A}\psi\|^2=\int_{-\infty}^{\infty} ds\left|\int_{-\infty}^{\infty} K(s, t)\psi(t)\, dt\right|^2=\int_{-\infty}^{\infty} ds\left|\int_{\mathrm{E}} \overline{K(s, t)}\overline{\psi(t)}\, dt\right|^2$$

$$=\int_{-\infty}^{\infty} ds\left|\int_{\mathrm{E}} w_s(t)\overline{\psi(t)}\, dt\right|^2=\int_{-\infty}^{\infty} |(w_s, \psi)|^2\, ds<\varepsilon^2.$$

We see that if the function $\psi(t)$ were also orthogonal to each of the functions $\varphi_k(t)$ $(k=1, 2, \ldots, m)$, then it would satisfy all the conditions of the lemma, and this lemma would be proved. So it remains to show how such a function can be constructed. To do this, we select $m+p+1$ linearly independent functions $\psi_k(t)\in \mathrm{L}^2$ which are equal to 0 outside E. Such functions exist, because E is of positive measure. (For example, we can select $m+p+1$ disjoint sub-sets of positive measure in E, and take the characteristic functions of these sub-sets as the functions $\psi_k(t)$.) Having selected the functions $\psi_k(t)$, we can put

$$\psi(t)=\sum_{k=1}^{m+p+1} \gamma_k\psi_k(t),$$

and demand that

$$\|\psi\| = 1, \qquad (\varphi_r, \psi) = 0 \qquad (r = 1, 2, \ldots, m)$$
$$(w_{s_k}, \psi) = 0 \qquad (k = 1, 2, \ldots, p).$$

This system of equations can be solved for the coefficients γ_k; in fact, these coefficients are determined by the equations essentially uniquely, that is, up to a constant factor equal to 1 in absolute value.

So the lemma is proved, and with it the proof of von Neumann's theorem is complete.

Appendix 2

Differential operators

§123. Self-adjoint differential expressions

This Appendix contains the main facts in the theory of ordinary differential operators. Treatments of various special topics and other approaches to the theory can be found in the monographs by E. C. Titchmarsh [124], B. M. Levitan [70], and M. A. Naimark [90]; see also Levitan and Sargsyan [72].

We begin by giving some facts about real-valued differential expressions which are self-adjoint in the Lagrange sense.

In textbooks on analysis it is shown that a self-adjoint, second-order differential expression

$$c_0D^2+c_1D+c_2D^0 \qquad \left(c_k=c_k(t),\ D^k=\frac{d^k}{dt^k}\right)$$

may be written, if its coefficient $c_k(t)$ is k times differentiable, in the form

$$-D(p_0D)+D^0p_1D^0,$$

where $p_0(t)$ is a differentiable function. It would, however, be possible to consider the differential expression $-D(p_0D)+D^0p_1D^0$ without assuming $p_0(t)$ to be differentiable, provided that this expression operated only on differentiable functions $\varphi(t)$ such that the product $p_0\varphi'$ is absolutely continuous.

We turn now to self-adjoint differential expressions of order $2n$. Here we take as the canonical form

$$l=p_nD^0-D\{p_{n-1}D-D[p_{n-2}D^2-\cdots-D(p_1D^{n-1}-Dp_0D^n)\cdots]\}. \quad (1)$$

If the coefficient $p_{n-k}(t)$ is differentiable k times $(k=0,1,\ldots,n)$, then (1) is an ordinary differential expression, and that it is self-adjoint can be verified immediately. Here, however, we shall not assume the differentiability of the coefficients $p_{n-k}(t)$, though we shall continue to call (1) a differential expression. Letting (a,b) denote the open interval (finite or infinite) on which the differential expression (1) is to be studied, we shall assume that

(i) the coefficients $p_k(t)$ are measurable in this interval (a, b), and
(ii) in every closed subinterval $[\alpha, \beta] \subset (a, b)$ they satisfy the conditions

$$\int_\alpha^\beta \frac{dt}{|p_0(t)|} < \infty, \qquad \int_\alpha^\beta |p_k(t)|\, dt < \infty \qquad (k = 1, 2, \ldots, n), \tag{2}$$

i.e., p_0^{-1} and p_k $(k = 1, 2, \ldots, n)$ are locally Lebesgue-integrable.

If the interval (a, b) is finite, and if the conditions (2) are satisfied when $\alpha = a$ and $\beta = b$, then the expression (1) is said to be *regular*.

If the interval (a, b) is infinite, or if it is finite but the conditions (2) do not hold when $\alpha = a$ or when $\beta = b$, then the expression (1) is said to be *singular*.

The end-point a is said to be *singular* if $a = -\infty$, or if $a > -\infty$ but the conditions (2) do not hold with $\alpha = a$. Singularity of the end-point b is defined similarly. If an end-point is not singular, then it is *regular*.

Without loss of generality we can regard singular differential expressions which have one singular end-point as being on the half-line $(0, \infty)$, and those which have two singular end-points as being on the whole line $(-\infty, \infty)$.

It is also convenient to introduce, instead of the derivatives $D^k\varphi = \varphi^{(k)}$, the quasi-derivatives $D^{[k]}\varphi = \varphi^{[k]}$ defined by

$$\begin{aligned} D^{[k]} &= D^k && (k = 0, 1, 2, \ldots, n-1),\\ D^{[n]} &= p_0 D^n, &&\\ D^{[n+k]} &= p_k D^{n-k} - DD^{[n+k-1]} && (k = 1, 2, \ldots, n). \end{aligned}$$

The expression (1) can then be written in the form

$$l = D^{[2n]}.$$

Let D^* denote the class of all those functions $\varphi(t) \in L^2(a, b)$ such that the quasi-derivatives $\varphi^{[k]}(t)$ $(k = 0, 1, \ldots, 2n-1)$ are absolutely continuous, and the quasi-derivative $\varphi^{[2n]}(t)$ belongs to $L^2(a, b)$. Clearly, D^* is the biggest linear manifold in $L^2(a, b)$ on which the expression l has a natural meaning and can be considered as an operator in $L^2(a, b)$. We denote this operator by L^*, and so $D^* = D(L^*)$. The suitability of this notation will be discovered later.

From the representation (1) it is easy to obtain *Lagrange's identity* for any pair of functions $\varphi, \psi \in D^*$:

$$l[\varphi]\bar{\psi} - \varphi l[\bar{\psi}] = \frac{d}{dt}[\varphi, \psi](t),$$

where $[\varphi, \psi](t)$ is the following bilinear form:

$$[\varphi, \psi](t) = \sum_{k=1}^{n} \{\varphi^{[k-1]}(t)\overline{\psi^{[2n-k]}(t)} - \varphi^{[2n-k]}(t)\overline{\psi^{[k-1]}(t)}\}.$$

We shall use Lagrange's identity in the form

$$\int_{\alpha}^{\beta} l[\varphi(t)]\overline{\psi(t)}\, dt - \int_{\alpha}^{\beta} \varphi(t) l[\overline{\psi(t)}]\, dt = [\varphi, \psi]_{\alpha}^{\beta},$$

where $[\alpha, \beta]$ is any closed sub-interval of the interval (a, b), and

$$[\varphi, \psi]_{\alpha}^{\beta} = [\varphi, \psi](\beta) - [\varphi, \psi](\alpha).$$

Since each of the integrals

$$\int_{\alpha}^{\beta} l[\varphi(t)]\overline{\psi(t)}\, dt \quad \text{and} \quad \int_{\alpha}^{\beta} \varphi(t) l[\overline{\psi(t)}]\, dt$$

exists when $\alpha = a$, $\beta = b$, it follows that:

the bilinear form $[\varphi, \psi](t)$ has a finite value at the end-points of the interval (a, b), independently of whether those end-points are regular or singular.

We take the value of the bilinear form at a singular end-point to be the limit of $[\varphi, \psi](t)$ as t tends to this end-point.

Let l be a differential expression, regular or singular, on an interval (a, b), and let $g(t)$ be a complex-valued function which is measurable and locally integrable on this interval. By a *solution* of the equation

$$l[y] - \lambda y = g(t) \tag{3}$$

for some complex value of λ it is natural to mean any function $\varphi(t)$ which, together with its quasi-derivatives up to order $(2n-1)$, is absolutely continuous and which makes this equation an identity almost everywhere in (a, b).

The following existence and uniqueness theorem holds for the equation (3); it is easily proved by Picard's method of successive approximations.

Theorem. *The equation* (3) *has a solution $\varphi(t)$ which satisfies the Cauchy-type conditions*

$$\varphi^{[k]}(t_0) = \alpha_k \qquad (\alpha_k = \text{const.};\ k = 0, 1, 2, \ldots, 2n-1),$$

where t_0 is any interior point of the interval (a, b) or a regular end-point, and this solution is unique.

We observe that, if the expression l is regular on the interval (a, b), and if $g(t) \in L^2(a, b)$, then, by the equation (3) itself, this solution will belong to D^*.

The existence theorem implies, in particular, that the linear manifold of solutions of the homogeneous equation

$$l[u]-\lambda u=0 \tag{4}$$

has dimension $2n$. Solutions $u_1, u_2, \ldots, u_{2n}$ of equation (4) are linearly independent if and only if their *Wronskian determinant*

$$W[u_1, u_2, \ldots, u_{2n}] \equiv \det [u_i^{[k-1]}(t)]_{i,k=1}^{2n}$$

is never equal to 0.

Any set of $2n$ linearly independent solutions of equation (4) is called a *fundamental system.*

If the functions

$$u_1(t), u_2(t), \ldots, u_{2n}(t) \tag{5}$$

form a fundamental system of solutions of equation (4), then any solution of the inhomogeneous equation (3) can be expressed in the form

$$\varphi(t)=\sum_{k=1}^{2n} u_k(t)\int_{t_0}^{t} v_k(s)g(s)\,ds+\sum_{k=1}^{2n} c_k u_k(t),$$

as may easily be shown by suitably modifying the classical method of 'variation of the parameters'. As in the classical case, it can be shown that the functions

$$v_1(t), v_2(t), \ldots, v_{2n}(t) \tag{6}$$

form another fundamental system of solutions of the equation (4), called the system *adjoint* to the system (5).

Later we shall need the following simple

Lemma. *If the differential expression l is regular on the interval (a, b), then the equation (3) will have a solution satisfying the boundary conditions*

$$y^{[k]}(a)=y^{[k]}(b)=0 \qquad (k=0, 1, 2, \ldots, 2n-1)$$

if and only if the right-hand side $g(t)$ is orthogonal to the $2n$-dimensional manifold of solutions of the homogeneous equation

$$l[u]-\bar{\lambda} u=0. \tag{4'}$$

Proof. Let $\varphi(t)$ be a solution of the equation (3) satisfying the conditions

$$\varphi^{[k]}(b)=0 \qquad (k=0, 1, 2, \ldots, 2n-1).$$

By applying Lagrange's identity to the function $\varphi(t)$ and some function $u_i(t)$ belonging to a fundamental system of solutions of the equation (4'), we obtain

$$\sum_{k=1}^{n} \{u_i^{[k-1]}(a)\overline{\varphi^{[2n-k]}(a)} - u_i^{[2n-k]}(a)\overline{\varphi^{[k-1]}(a)}\} = \int_a^b u_i(s)\overline{g(s)}\, ds. \quad (7)$$

By subjecting the fundamental system to the initial conditions

$$u_i^{[k-1]}(a) = \delta_{ik} = \begin{cases} 0 & (i \neq k), \\ 1 & (i = k), \end{cases}$$

we obtain

$$\overline{\varphi^{[r]}(a)} = \begin{cases} -\int_a^b u_{2n-r}(s)\overline{g(s)}\, ds & (r = 0, 1, 2, \ldots, n-1), \\ \int_a^b u_{2n-r}(s)\overline{g(s)}\, ds & (r = n, n+1, \ldots, 2n-1), \end{cases} \quad (8)$$

from which the lemma follows.

Up to now in this section we have been considering self-adjoint differential expressions with *real* coefficients. Such expressions are necessarily of even order (we are talking about scalar differential expressions, not about systems of differential equations). In the general case, self-adjoint differential expressions with *complex* coefficients may be of any order $r \geqslant 1$, either even ($2n$) or odd ($2n+1$). If the coefficients are sufficiently smooth, these self-adjoint expressions can be written in the form

$$l = \sum_{k=0}^{r} i^k l_k \qquad (i^2 = -1),$$

where

$$l_{2j} = D^j(p_{n-j}D^j),$$

$$l_{2j+1} = \tfrac{1}{2}D^j\{Dq_{n-j} + \bar{q}_{n-j}D\}D^j,$$

and $n = \left[\frac{r}{2}\right]$ and the functions $p_k(t)$ are real-valued.

The expression l above can be put into a form analogous to (1), and we can also introduce for it quasi-derivatives (see F. S. Rofe-Beketov [108] and Hartman [43]), as we did earlier; this enables smoothness conditions on the coefficients to be dispensed with, except that, for an expression of odd order, $q_0(t)$ must be locally absolutely continuous.

We point out, however, that everywhere in this Appendix, we assume the coefficients of the differential expressions to be real-valued, unless otherwise stated.

§124. Regular differential operators

Let l be a self-adjoint, regular, differential expression, defined on an interval (a, b). If φ, ψ are any two functions chosen from D^*, then, in general, the difference

$$(L^*\varphi, \psi) - (\varphi, L^*\psi) = [\varphi, \psi]_a^b$$

is not equal to 0, and therefore L^* is not a symmetric operator. To make the right-hand side equal to 0, some extra conditions must be imposed on φ and ψ, and thus we have to go from the domain D^* to some sub-set of it. It is in every case sufficient to require that each of the functions φ, ψ shall satisfy the relations

$$y^{[k]}(a) = y^{[k]}(b) = 0 \qquad (k = 0, 1, 2, \ldots, 2n-1). \tag{1}$$

Let D denote the set of all functions belonging to D^* which satisfy the $4n$ conditions (1). It is natural to expect that the operator L which has its domain $D(L) = D$ and which coincides with L^* on this domain will be a symmetric operator. And since, for all $\varphi, \psi \in D$,

$$(L\varphi, \psi) = (\varphi, L\psi),$$

it is in fact only necessary to show that the manifold D is dense in $L^2(a, b)$ to justify our expectation. This fact is obvious when l is a differential expression in the ordinary sense, because in that case the manifold D contains, for example, all polynomials which satisfy the conditions (1). To prove that D is dense in $L^2(a, b)$ in the general case, we observe that, by the lemma in §123, the space $L^2(a, b)$ may be resolved into the orthogonal sum

$$L^2(a, b) = R(L) \oplus N_0, \tag{2}$$

where N_0 is the $2n$-dimensional manifold of solutions of the homogeneous equation

$$l(u) = 0.$$

We now suppose that $(h, \varphi) = 0$ for some $h \in L^2(a, b)$ and for all $\varphi \in D$, and want to show that this implies that $h = 0$. To do this, let ψ be any solution of the equation

$$l[\psi] = h.$$

By Lagrange's identity, we have

$$(\psi, l[\varphi]) = (l[\psi], \varphi) = (h, \varphi) = 0.$$

Since $l[\varphi]=L^*\varphi=L\varphi\in \mathrm{R}(L)$, it follows from (2) that $\psi\in \mathrm{N}_0$, *i.e.*,

$$l[\psi]=0,$$

and this implies that $h=0$, as required.

We now prove that the operator L^* is adjoint to L, and so justify the notation. For the moment, let M be the operator adjoint to L. Let $\psi\in \mathrm{D}(M)$ and let

$$M\psi=\chi.$$

Let ψ_0 be a solution of the equation

$$l[y]=\chi.$$

Then, by Lagrange's identity, for any $\varphi\in \mathrm{D}$,

$$(\varphi, \chi)=(\varphi, l[\psi_0])=(l[\varphi], \psi_0)=(L\varphi, \psi_0).$$

But by the definition of the adjoint operator,

$$(\varphi, \chi)=(\varphi, M\psi)=(L\varphi, \psi).$$

Thus

$$(L\varphi, \psi-\psi_0)=0.$$

Hence, since $\varphi\in \mathrm{D}(L)$ is arbitrary, it follows, by (2), that

$$\psi-\psi_0\in \mathrm{N}_0.$$

This inclusion shows that $\psi\in \mathrm{D}^*$ and consequently that

$$M\psi=\chi=l[\psi_0]=l[\psi].$$

We have thus shown that

$$M\subseteq L^*.$$

On the other hand, for any $\varphi\in \mathrm{D}$ and any $\psi\in \mathrm{D}^*$,

$$(L\varphi, \psi)-(\varphi, L^*\psi)=[\varphi, \psi]_a^b=0.$$

Hence $L^*\subseteq M$.

Therefore $M=L^*$, as asserted.

The reader may easily verify that L^{**} coincides with L, and this implies that L is closed.

As already mentioned in §123, $\mathrm{D}(L^*)$ is the biggest linear manifold on which the differential expression l generates an operator in $\mathrm{L}^2(a, b)$. Every closed symmetric operator which is generated by the expression l and for which the domain of the adjoint operator lies in $\mathrm{D}(L^*)$ is an extension of L. So we shall call L the *regular differential operator with the* *minimal domain*, or briefly, the *minimal differential operator* (generated by the differential expression l).

Since, for any λ, the equation

$$L^*\psi - \lambda\psi = 0$$

has $2n$ linearly independent solutions, the deficiency index of the operator L is $(2n, 2n)$.

To describe all the self-adjoint extensions of an operator L, it suffices to indicate the domains of all these extensions.† We shall show that the domain of any self-adjoint extension of an operator L can be specified by means of particular boundary conditions, and we shall give the nature of all such conditions.

We point out first of all that, from Lagrange's identity and the lemma stated at the end of §46 we obtain the following

Criterion for self-adjointness.

If $\tilde{L}$ is an extension of the operator L, then the operator $\tilde{L}$ will be self-adjoint if and only if the following two conditions hold:

1) $\tilde{L} \subset L^*$,
2) *the inclusion $\varphi \in \mathrm{D}$ is equivalent to the equality*

$$[\varphi, \psi]_a^b = 0 \tag{3}$$

for all $\psi \in \mathrm{D}(\tilde{L})$.

Our first task will be to transform this criterion into a more convenient form. We deal with this problem in the next section, which is based on one of F. S. Rofe-Beketov's papers (see [109]).

§125. Self-adjoint extensions of a regular differential operator

If with each function $\varphi \in \mathrm{D}(L^*)$ there is associated an ordered pair $\{\hat{\varphi}, \hat{\varphi}'\}$ of vectors of the $2n$-dimensional, complex, Euclidean space $E_{2n} = E$ (see §5) in accordance with the rule

$$\hat{\varphi} = \{\varphi(a), \varphi'(a), \ldots, \varphi^{(n-1)}(a), \varphi(b), \ldots, \varphi^{(n-1)}(b)\}, \tag{1}$$

$$\hat{\varphi}' = \{\varphi^{[2n-1]}(a), \ldots, \varphi^{[n]}(a), -\varphi^{[2n-1]}(b), \ldots, -\varphi^{[n]}(b)\}, \tag{1'}$$

† Here and later, we mean by 'self-adjoint extensions' self-adjoint extensions of the first kind.

For the general form of boundary conditions which characterize self-adjoint extensions of differential operators, see M. G. Krein [60] and Graff [41].

then, for any φ, ψ in $D(L^*)$,

$$[\varphi, \psi]_a^b = (\hat{\varphi}', \hat{\psi})_E - (\hat{\varphi}, \hat{\psi}')_E,$$

where $(\cdot, \cdot)_E$ denotes the scalar product in E.

By virtue of this observation we can replace the conditions 1) and 2) in the criterion for self-adjointness stated in §124 by the following conditions:

1) $\tilde{L} \subset L^*$,
2) *the inclusion* $\varphi \in D(\tilde{L})$ *is equivalent to the equality*

$$(\hat{\varphi}', \hat{\psi})_E = (\hat{\varphi}, \hat{\psi}')_E \tag{2}$$

for all $\psi \in D(\tilde{L})$.

It is convenient to formulate this result by means of the concept of binary relations. We explain this concept in its general form.

Let M be a given set, and suppose that a rule is given by means of which ordered pairs of elements x, y are selected from M. This rule, or, equivalently, the sub-set in $M \times M$ of the ordered pairs $\{x, y\}$ is called a *binary relation*† in M. This sub-set, or the rule mentioned, is denoted by a single letter, say $\mathbf{r}$, and we write $x\mathbf{r}y$ as well as $\{x, y\} \in \mathbf{r}$.

In the question in which we are interested, the formulae (1), (1′) determine in E a relation $\mathbf{r}$ whose elements are the ordered pairs $\{\hat{\varphi}, \hat{\psi}'\}$ having the property (2).

Definition. A binary relation $\mathbf{r}$ defined in E will be called an *Hermitian relation* if it has the following properties:

1°. if $x\mathbf{r}x'$ and $y\mathbf{r}y'$, then

$$(x', y)_E = (x, y')_E; \tag{3}$$

2°. if (3) holds for certain elements x and x' in E and for all pairs $\{y, y'\} \in \mathbf{r}$, then $\{x, x'\} \in \mathbf{r}$.

It is easy to show that an Hermitian relation is linear, *i.e.*, if $x\mathbf{r}x'$ and $y\mathbf{r}y'$, then, for any (complex) α, β,

$$(\alpha x + \beta y)\mathbf{r}(\alpha x' + \beta y').$$

We now give a compact formulation of the criterion stated at the beginning of this section.

Theorem 1. *A linear manifold* $\tilde{D} \subset D(L^*)$ *will be the domain* $D(\tilde{L})$ *of some self-adjoint extension* $\tilde{L}$ *of an operator* L *if and only if there is an*

† See, *e.g.*, Maclane [80]. In this book M is a vector space, and a relation is taken to be, by definition, a linear manifold in $M \times M$.

Hermitian relation $\mathbf{r}$ *in the* $2n$*-dimensional Euclidean space* E *such that*

$$\varphi \in \tilde{\mathrm{D}} \Longleftrightarrow \hat{\varphi}\mathbf{r}\hat{\varphi}', \tag{4}$$

where $\hat{\varphi}, \hat{\varphi}' \in \mathrm{E}$ *are connected with* $\varphi \in \mathrm{D}(L^*)$ *by the formulae* (1), (1').†

We note that, if $v\mathbf{r}v'$ and $v' = iv$, then $v = 0$. For, by definition of an Hermitian relation,

$$(v, v')_{\mathrm{E}} = (v', v)_{\mathrm{E}};$$

but in this case the left-hand side is $-i(v, v)_{\mathrm{E}}$, and the right-hand side is $i(v, v)_{\mathrm{E}}$.

From this observation we obtain the following propositions:

A. *An ordered pair* $\{x, x'\} \in \mathbf{r}$ *is uniquely determined by each of the vectors*

$$x^+ = x + ix' \quad \text{and} \quad x^- = x - ix'.$$

B. *When* $x\mathbf{r}x'$ *runs through the whole relation* $\mathbf{r}$, *each of the vectors* x^+ *and* x^- *runs through the whole space* E.

C. *The operator* $U = U_{\mathbf{r}}$ *defined by the formula*

$$U(x + ix') = x - ix',$$

where $x\mathbf{r}x'$, *is unitary.*

Clearly, we need only prove the isometry of the operator U. This is the proof:

$$\begin{aligned} \|x \pm ix'\|_{\mathrm{E}}^2 &= \|x\|_{\mathrm{E}}^2 + \|x'\|_{\mathrm{E}}^2 \pm [(x, ix')_{\mathrm{E}} + (ix', x)_{\mathrm{E}}] \\ &= \|x\|_{\mathrm{E}}^2 + \|x'\|_{\mathrm{E}}^2, \end{aligned}$$

because the relation (3) implies that, when $\{x, x'\} = \{y, y'\} \in \mathbf{r}$,

$$(x', x)_{\mathrm{E}} - (x, x')_{\mathrm{E}} = 0.$$

The operator U is naturally called the *Cayley transform* of the Hermitian relation $\mathbf{r}$.

Theorem 2. *For any self-adjoint operator* A *and unitary operator* U *in the space* E, *the equations*

$$(\cos A)x - (\sin A)x' = 0, \tag{5}$$

$$(U - I)x + i(U + I)x' = 0 \tag{6}$$

† It is clear that, not only can a $\hat{\varphi}$ and $\hat{\varphi}'$ be found corresponding to each $\varphi \in \mathrm{D}(L^*)$, but also, for any $\hat{\varphi}, \hat{\varphi}'$ in E, a function $\varphi \in \mathrm{D}(L^*)$ can be found (not uniquely, of course) so that (1) and (1') are satisfied.

each determine an Hermitian relation; and conversely, every Hermitian relation $\mathbf{r}$ *can be expressed in the forms* (5) *and* (6). *Further, the operators* $-\exp[2iA]$ *and* U *in the relations* (5), (6) *are uniquely determined by the relation* $\mathbf{r}$ *and are its Cayley transform.*

Proof. We observe first of all that the relation $\mathbf{r}$ given by the equation (5) can be written in the form

$$x = (\sin A)v, \qquad x' = (\cos A)v, \tag{5'}$$

and the element $v \in E$ is uniquely determined by the ordered pair $x\mathbf{r}x'$, since (5′) implies that

$$v = (\sin A)x + (\cos A)x',$$

and conversely, all elements $x\mathbf{r}x'$ determined by the equation (5) can be obtained by means of the parametric representation (5′) when v runs through E.

From what has been said, we can verify that the relation $\mathbf{r}$ given by the equation (5) is an Hermitian relation. For, if x, x' and y, y' satisfy (5), then in addition to the representation (5′) we also have the representation

$$y = (\sin A)w, \qquad y' = (\cos A)w, \tag{5''}$$

where w is some element of E. But then

$$(x, y')_E = ((\sin A)v, (\cos A)w)_E = (v, (\sin A \cos A)w)_E$$

and

$$(x', y)_E = ((\cos A)v, (\sin A)w)_E = (v, (\cos A \sin A)w)_E$$
$$= (v, (\sin A \cos A)w)_E,$$

i.e.,

$$(x, y')_E = (x', y)_E$$

and so condition 1° in the definition of an Hermitian relation holds. To verify condition 2°, we select 2 elements $x, x' \in E$ and assume that, for any pair $y\mathbf{r}y'$,

$$(x, y')_E = (x', y)_E.$$

This implies that, for any $w \in E$,

$$(x, (\cos A)w)_E = (x', (\sin A)w)_E$$

or

$$((\cos A)x - (\sin A)x', w)_E = 0,$$

from which it follows that

$$(\cos A)x - (\sin A)x' = 0, \qquad \textit{i.e.,} \quad x\mathbf{r}x'.$$

The relation $\mathbf{r}$ given by the equation (6) reduces to the one already discussed, if we put $U = -\exp(2iA)$.

Thus one part of Theorem 2 has been proved.

As regards the second part, it suffices to prove it as applied to equation (6). If **r** is a given Hermitian relation, then from proposition C above it follows that the equation (6) holds in every case when $U = U_{\mathbf{r}}$, where $U_{\mathbf{r}}$ is the Cayley transform of the Hermitian relation **r**. But by propositions A and B this equation will hold only when $U = U_{\mathbf{r}}$.

From Theorems 1 and 2 we obtain the main

Theorem 3. *Every self-adjoint extension $\tilde{L}$ of a minimal operator L is determined by boundary conditions of the form either*

$$(\cos A)\hat{\varphi} - (\sin A)\hat{\varphi}' = 0, \tag{7}$$

or

$$(U - I)\hat{\varphi} + i(U + I)\hat{\varphi}' = 0, \tag{8}$$

where A is an Hermitian, and U a unitary, $2n \times 2n$ matrix, and the vectors $\hat{\varphi}$, $\hat{\varphi}'$ are determined from $\varphi(t) \in D(L^*)$ by the formulae (1), (1′).

Conversely, any boundary condition of the form (7) or (8) determines a self-adjoint extension $\tilde{L}$ of the operator L.

Boundary conditions which generate self-adjoint extensions of a differential operator are said to be *self-adjoint.*

Examples. 1. *Separated, self-adjoint boundary conditions* for an operator of order $2n$ reduce to the form

$$(\cos A_a)\hat{\varphi}_a - (\sin A_a)\hat{\varphi}'_a = 0,$$

$$(\cos A_b)\hat{\varphi}_b + (\sin A_b)\hat{\varphi}'_b = 0,$$

where A_a and A_b are arbitrary, Hermitian $n \times n$ matrices, and the vectors $\hat{\varphi}_t, \hat{\varphi}'_t$ $(t = a, b)$ are defined by the formulae

$$\hat{\varphi}_t = \{\varphi(t), \varphi'(t), \ldots, \varphi^{(n-1)}(t)\},$$

$$\hat{\varphi}'_t = \{\varphi^{[2n-1]}(t), \varphi^{[2n-2]}(t), \ldots, \varphi^{[n]}(t)\}.$$

These conditions are obtained from (7) if we take as A the block-diagonal matrix

$$A = \begin{bmatrix} A_a & 0 \\ 0 & A_b \end{bmatrix}.$$

In particular, if A is a diagonal matrix, then we obtain self-adjoint conditions, of the form

$$\left.\begin{aligned} \sin \alpha_k \,.\, \varphi^{[2n-k]}(a) - \cos \alpha_k \,.\, \varphi^{(k-1)}(a) = 0 \\ \sin \beta_k \,.\, \varphi^{[2n-k]}(b) + \cos \beta_k \,.\, \varphi^{(k-1)}(b) = 0 \end{aligned}\right\} \qquad (k = 1, 2, \ldots, n),$$

where α_k, β_k are arbitrary real parameters.

2. *Generalized periodic conditions*

$$\left.\begin{aligned} \exp(i\alpha_k)\varphi^{[2n-k]}(b) &= \varphi^{[2n-k]}(a) \\ \exp(i\alpha_k)\varphi^{(k-1)}(b) &= \varphi^{(k-1)}(a) \end{aligned}\right\} \qquad (k=1,2,\ldots,n),$$

where the α_k are arbitrary real numbers.

These conditions are obtained from (8), (1), (1′) when

$$U=\left[\begin{array}{ccc|ccc} & & & \exp(i\alpha_1) & & \\ & 0 & & & \ddots & \\ & & & & & \exp(i\alpha_n) \\ \hline \exp(-i\alpha_1) & & & & & \\ & \ddots & & & 0 & \\ & & \exp(-i\alpha_n) & & & \end{array}\right]$$

These boundary conditions are transformed into equivalent conditions if the equalities (7) or (8) are multiplied on the left by an arbitrary non-singular $2n\times 2n$ matrix. So the question arises: under what circumstances will a condition of the form

$$B\hat{\varphi}-C\hat{\varphi}'=0, \tag{9}$$

where B, C are certain $2n\times 2n$ matrices, be self-adjoint? The answer to this question is given by

Theorem 4. *The boundary condition* (9) *will generate a self-adjoint extension of the operator* L *if and only if*

$$\det[BB^*+CC^*]\neq 0 \tag{10}$$

and

$$BC^*=CB^*. \tag{11}$$

The equation (9) with just the one condition (10) determines a quasi-self-adjoint extension of the operator L in the sense of §114.

For the proof of this theorem, and for other interesting facts, the reader is referred to Rofe-Beketov [109].

By carrying out the usual calculations, we can easily construct the Green's function $\tilde{G}(t,s)$ for any self-adjoint extension $\tilde{L}$, of the operator L, which does not have $\lambda=0$ as an eigenvalue. The function $\tilde{G}(t,s)$ is a Hilbert–Schmidt kernel, and the integral operator $\tilde{K}$ which is defined by this kernel is related to the operator L by the formula

$$\tilde{K} = \tilde{L}^{-1}.$$

Hence it follows that:

the spectrum of the operator $\tilde{L}$ consists only of eigenvalues, and these have just one limit-point, namely, at infinity.

The spectrum of $\tilde{L}$ will be as stated, even if $\lambda = 0$ is an eigenvalue. For it is always possible, by changing the parameters in the boundary conditions, to find a self-adjoint extension $\tilde{L}_1$ of the operator L such that 0 is not an eigenvalue, and therefore the spectrum of $\tilde{L}_1$ will have the stated properties. But then these same properties will be preserved for the spectrum of *any* self-adjoint extension $\tilde{L}$ of the operator L (because the extension is finite-dimensional).

§126. Singular differential operators†

We begin with the case where the interval is $(0, \infty)$, *i.e.*, the right-hand end-point is singular, and the left-hand end-point is taken to be regular.

We define an operator $\mathring{L}$ by the equality

$$\mathring{L}\varphi = l[\varphi]$$

on the manifold $D(\mathring{L})$ defined as the set of those functions $\varphi \in D^*$ which have compact support and satisfy the conditions

$$\varphi(0) = \varphi^{[1]}(0) = \cdots = \varphi^{[2n-1]}(0) = 0. \tag{1}$$

Since the operator $\mathring{L}$ is evidently symmetric, it admits closure; we denote its closure by L, and call it (see §124) the *singular differential operator* with the *minimum domain*, or, briefly, the *minimal differential operator.*

Theorem 1. The operator L^* is the adjoint of the operator L.

Proof. As in §124, we let M denote the operator adjoint to L. Clearly,

$$D(L^*) \subseteq D(M)$$

and, for $\psi \in D(L^*)$,

$$M\psi = l[\psi];$$

† §§126–128 are taken from I. M. Glazman's paper [36].

so we need only prove that

$$D(M) \subseteq D(L^*).$$

To do this, we choose some function $\psi \in D(M)$, and, as in §124, we put

$$M\psi = \chi.$$

Further, we let ψ_0 denote some solution of the equation

$$l[y] = \chi.$$

It now remains for us to prove that

$$\psi(t) = \psi_0(t) + u(t),$$

where $u(t)$ is some solution of the equation

$$l[y] = 0. \tag{2}$$

For this purpose we select a function $g(t) \in L^2(0, \infty)$ which is equal to 0 for $t \geqslant a$ and is orthogonal to the $2n$-dimensional manifold of solutions of equation (2) in the interval $0 \leqslant t \leqslant a$. Then, by §123, there is a function $\varphi(t)$ which is equal to 0 for $t \geqslant a$, and which satisfies the equation

$$l[\varphi] = g,$$

and also the condition (1). It is then clear that $\varphi \in D(\mathring{L}) \subset D(L)$, and therefore

$$(g, \psi) = (L\varphi, \psi) = (\varphi, M\psi) = (\varphi, \chi).$$

But

$$(g, \psi_0) = (l[\varphi], \psi_0) = (\varphi, l[\psi_0]) = (\varphi, \chi).$$

Therefore

$$(g, \psi - \psi_0) = 0,$$

from which, in view of the extent to which the choice of $g(t)$ is arbitrary, we conclude that

$$\psi(t) - \psi_0(t) = u(t),$$

where $u(t)$ is a solution of the equation (2). Q.E.D.

Remark 1. In the above proof, although $\psi_0 \notin L^2(0, \infty)$, the expression (g, ψ_0) makes sense because g is a function with compact support.

Remark 2. We note that every function in $D(L)$ satisfies the condition (1). For, let $\psi(t)$ be an arbitrary function in $D(L^*)$ which is equal to 0 for

$t \geqslant a$. Then, for $\varphi \in \mathrm{D}(L)$, the equalities

$$(L\varphi, \psi) = (\varphi, L^*\psi),$$

$$(l[\varphi], \psi) = [\varphi, \psi]_0^a + (\varphi, l[\psi])$$

imply that

$$[\varphi, \psi]_0^a = 0,$$

and so

$$[\varphi, \psi](0) = 0;$$

since the values $\psi^{[k-1]}(0)$, $k = 1, 2, \ldots, 2n$ are arbitrary, this implies that the function $\varphi(t)$ satisfies the condition (1).

We pass on now to the question of the deficiency numbers of the operator L. Since the coefficients of the differential expression l are real, the deficiency numbers are equal.

Theorem 2. The deficiency number m of a differential equation of order $2n$ on an interval with one regular end-point satisfies the inequality

$$n \leqslant m \leqslant 2n. \tag{3}$$

Proof. The right-hand inequality follows immediately from Theorem 1.

To prove the left-hand inequality, we use the first Neumann formula (§102):

$$\mathrm{D}(L^*) - \mathrm{D}(L) \oplus \mathrm{N}_\lambda \oplus \mathrm{N}_{\bar{\lambda}}.$$

Since

$$\dim \mathrm{N}_\lambda = \dim \mathrm{N}_{\bar{\lambda}},$$

it suffices to show that

$$\dim \mathrm{D}(L^*) \geqslant 2n \qquad (\mathrm{mod}\ \mathrm{D}(L)).$$

To do this, we show that there are $2n$ linearly independent functions belonging to $\mathrm{D}(L^*)$ and such that no non-trivial linear combination of them belongs to $\mathrm{D}(L)$. It is easy to see that we may take as these functions any $2n$ linearly independent functions $\psi_1(t), \ldots, \psi_{2n}(t)$ belonging to $\mathrm{D}(L^*)$ which satisfy the condition

$$\det\, [\psi_i^{[k-1]}(0)]_{i,k=1}^{2n} \neq 0,$$

for then the hypothesis

$$\psi = \sum_{i=1}^{2n} \alpha_i \psi_i \in \mathrm{D}(L)$$

implies (see Remark 2 after Theorem 1) the equalities

$$\psi(0) = \psi^{[1]}(0) = \cdots = \psi^{[2n-1]}(0) = 0,$$

from which it follows that all the $\alpha_i = 0 \quad (i = 1, 2, \ldots, 2n)$. Q.E.D.

From Theorems 1 and 2 it follows that the number of solutions of the equation

$$l[u] - \lambda u = 0 \qquad (\mathscr{I}(\lambda) \neq 0)$$

which belong to $L^2(0, \infty)$ is not less than half the order of the differential expression l (and does not depend on λ).

The integer m in the inequality (3) can take any value between n and $2n$. It is curious to note that, if

$$l_1 = -i(1+t)D(1+t) \qquad \text{and} \qquad l_2 = D^2 - 1,$$

then the operator of order $2n$, with minimal domain, which is generated by the differential expression

$$l = l_1^{m-n} l_2^{2n-m} l_1^{m-n} \qquad (n \leqslant m \leqslant 2n)$$

on the interval $(0, \infty)$ has the deficiency index (m, m). The verification of this fact is left to the reader; see Glazman [36].

The treatment in the case of the interval $(-\infty, \infty)$ is similar to the above.

The operator L with minimal domain is again defined as the closure of the operator $\mathring{L}$, but the condition (1) has to be dropped. Theorem 1 characterizing the operator L^* is the same, but Theorem 2 is replaced by

Theorem 3. *Let $L^{(-)}$ and $L^{(+)}$ be the singular differential operators, with minimal domains, generated by the differential expression l in the intervals $(-\infty, 0)$ and $(0, \infty)$ respectively. Then the deficiency number m of the operator L is given by*

$$m = m^{(-)} + m^{(+)} - 2n,$$

where $m^{(-)}$ and $m^{(+)}$ are the deficiency numbers of the operators $L^{(-)}$ and $L^{(+)}$ respectively.

Proof. We contract the domain of the operator $\mathring{L}$ by subjecting the functions φ belonging to the domain $D(\mathring{L})$ to the additional requirement that

$$\varphi(0)=\varphi^{[1]}(0)=\cdots=\varphi^{[2n-1]}(0)=0,$$

and we denote the linear manifold obtained in this way by D_0, and let $\mathring{L}_0$ denote the operator which coincides with $\mathring{L}$ on D_0 but is defined only on D_0. Let L_0 be the closure of $\mathring{L}_0$. Then it is clear that

$$\dim D(L)=2n \qquad (\operatorname{mod} D(L_0)).$$

But by a remark in §102 (p. 356), $\dim D(L) \quad (\operatorname{mod} D(L_0))$ is equal to the excess of the deficiency number of the operator L_0 over the deficiency number of the operator L. Therefore the deficiency number of the operator L_0 is equal to

$$m_0=m+2n.$$

On the other hand, the operator L_0 is reduced by the sub-spaces $L^2(-\infty, 0)$ and $L^2(0, \infty)$ and

$$L_0=L^{(-)}\oplus L^{(+)},$$

and so

$$m_0=m^{(-)}+m^{(+)}.$$

Thus

$$m=m^{(-)}+m^{(+)}-2n,$$

as we had to show.

In contrast to symmetric differential operators with real coefficients, the deficiency numbers of a singular symmetric differential operator with complex coefficients can also be unequal to one another, the simplest example of which is the operator $P=i(d/dt)$ on the half-line $[0, \infty)$ (see §102). The deficiency numbers of an operator of even order also can be unequal.

Theorem 4. (See W. N. Everitt [26]).
The deficiency numbers (m, s) of the minimal symmetric differential operator L of arbitrary order $r=2n$ or $r=2n+1$ with complex coefficients on the half-line satisfy the inequalities

$$\left[\frac{r}{2}\right]\leqslant m, s\leqslant r; \quad m+s\geqslant r.$$

(The last inequality shows that if one of the deficiency numbers of an operator of order $r=2n+1$ is n, then the other is not less than $n+1$.)

Theorem 5.† *If one of the deficiency numbers (m, s) of a symmetric differential operator L of order $r > 1$ is equal to r, then the other is also equal to r, i.e., $m = s = r$.*

(When $r = 1$, this theorem is false, as the example of the operator $i(d/dt)$ on the half-line shows.)

It can be proved‡ that, for symmetric differential operators with complex coefficients, the deficiency numbers can take *any* values which are compatible with Theorems 4 and 5 and which satisfy the additional restriction that $|m - s| \leq 1$.

§127. Self-adjoint extensions of a singular differential operator

Self-adjoint extensions of a singular differential operator, like those of a regular operator, can be characterized by means of a set of boundary conditions, though these will have a more complicated structure than the corresponding conditions in §125 had. We deal with the case where just one end-point of the interval is singular. From the general theory of extensions (see §101) we obtain the following

Theorem 1. Let the deficiency index of the operator L be (m, m) $(n \leq m \leq 2n)$, and let the functions

$$u_k(t; \lambda) \in L^2(0, \infty) \qquad (k = 1, 2, \ldots, m)$$

form an orthonormal set of solutions of the equation

$$l[u] - \lambda u = 0$$

for some fixed, non-real value of λ.

Further, if Γ_θ is the linear envelope of the functions

$$w_i^{(\theta)}(t; \lambda) = u_i(t; \lambda) + \sum_{k=1}^{m} \theta_{ik} \overline{u_k(t; \lambda)},$$

where $\theta = [\theta_{ik}]$ is a unitary matrix of the mth order, then the formula

$$D(L_\theta) = D(L) \oplus \Gamma_\theta \tag{1}$$

defines a one-to-one correspondence between the class of all self-adjoint extensions L_θ of the operator L and the class of all unitary matrices θ of order m.

† For operators with real coefficients this theorem is trivial. For operators of even order with complex coefficients it was established by W. N. Everitt [27]. For operators of odd order, see V. I. Kogan and F. S. Rofe-Beketov [53].

‡ See V. I. Kogan and F. S. Rofe-Beketov [54].

Proof. It suffices to observe that the functions

$$\overline{u_k(t;\lambda)} \qquad (k=1,2,\ldots,m)$$

form an orthonormal set of solutions of the equation

$$L^*g-\bar{\lambda}g=0. \qquad \text{Q.E.D.}$$

What boundary conditions characterize the functions in $D(L_\theta)$? That is the next question to be investigated.

Theorem 2. Let L_θ be a self-adjoint extension of an operator L with deficiency index (m, m) defined, in accordance with Theorem 1, by a unitary matrix θ, and let the functions $w_i^{(\theta)}(t;\lambda)$ $(i=1,2,\ldots,m)$ be as in Theorem 1.

Then the domain $D(L_\theta)$ of the operator L_θ consists of all those functions $\varphi(t)\in D(L^*)$ which satisfy the m conditions

$$[\varphi, w_i^{(\theta)}]_0^\infty=0 \qquad (i=1,2,\ldots,m). \tag{2}$$

Proof. First of all we note that, by the Lagrange formula, a function $\varphi(t)\in D(L^*)$ will belong to $D(L_\theta)$ if and only if, for all $\psi(t)\in D(L_\theta)$,

$$[\varphi,\psi]_0^\infty=0, \tag{3}$$

It now remains to show that the set of conditions (3), which contains an 'arbitrary' function $\psi(t)$, reduces (when $\varphi(t)\in D(L^*)$) to the set of m conditions (2).

By formula (1), the function $\psi(t)$ admits the representation

$$\psi(t)=\varphi_0(t)+\sum_{i=1}^m c_i w_i^{(\theta)}(t;\lambda) \qquad (\varphi_0(t)\in D(L)),$$

and so

$$[\varphi,\psi]_0^\infty=[\varphi,\varphi_0]_0^\infty+\sum_{i=1}^m c_i[\varphi, w_i^{(\theta)}]_0^\infty.$$

Since $\varphi_0(t)\in D(L)$,

$$[\varphi,\varphi_0]_0^\infty=0 \qquad \text{for} \qquad \varphi(t)\in D(L^*),$$

and since the constants c_i $(i=1,2,\ldots,m)$ are arbitrary, it follows that the conditions (3) and (2) are equivalent. Q.E.D.

The equalities (2) entail a passage to the limit, but we still call them boundary conditions. The next theorem shows a case where the boundary conditions do not entail a passage to the limit.

Theorem 3. *If the deficiency index of the operator L is (n, n), then the set of boundary conditions defining a self-adjoint extension L_θ of the operator L has the form*

$$[\varphi, w_i^{(\theta)}](0)=0 \qquad (i=1, 2, \ldots, n) \tag{4}$$

Proof. We have only to show that under the conditions of the theorem any two functions $\varphi(t)$, $\psi(t)$ belonging to $D(L^*)$ satisfy the relation

$$[\varphi, \psi](\infty)=0.$$

To do this, we choose functions

$$\psi_i(t) \qquad (i=1, 2, \ldots, 2n), \qquad \psi_i(t)=0 \quad \text{for} \quad t>a>0,$$

belonging to $D(L^*)$ and such that

$$\det\,[\psi_i^{[k-1]}(0)]_{i,k=1}^{2n}\neq 0.$$

For such a choice of the functions $\psi_i(t)$ their linear envelope and the domain $D(L)$ have no non-zero elements in common. On the other hand, since the deficiency index of the operator L is (n, n),

$$\dim D(L^*)=2n \qquad (\text{mod}\, D(L)),$$

and therefore, for any function $\varphi(t)\in D(L^*)$, there are constants $\alpha_1, \alpha_2, \ldots, \alpha_{2n}$ such that

$$\varphi(t)-\sum_{i=1}^{2n}\alpha_i\psi_i(t)=\varphi_0(t)\in D(L).$$

Hence, since the functions $\psi_i(t)$ $(i=1, 2, \ldots, 2n)$ vanish for $t>a$, and since

$$\varphi_0^{[k-1]}(0)=0 \qquad (k=1, 2, \ldots, 2n),$$

we find that

$$[\varphi, \psi](\infty)=\left[\left(\varphi_0+\sum_{i=1}^{2n}\alpha_i\psi_i\right), \psi\right](\infty)=[\varphi_0, \psi](\infty)=0. \qquad \text{Q.E.D.}$$

It should be noted, that in the general case, the boundary conditions depend essentially on the functions

$$w_i^{(\theta)} \qquad (i=1, 2, \ldots, m)$$

and therefore they cannot be written down until the operator L has been specified. But if the deficiency index of the operator L is (n, n), then, by the method described in §125, the boundary conditions (4) can be freed from the presence of the functions $w_i^{(\theta)}(t; \lambda)$, and therefore they do not depend on L.

If the deficiency index is $(2n, 2n)$, then the number of boundary conditions to define a self-adjoint extension of a singular operator is $2n$, just as in the case of a regular operator. For this and other reasons (see Theorem 2 in the next section), we shall call a singular operator with deficiency index $(2n, 2n)$ *quasi-regular.*

For a second-order differential operator there are only two possibilities, either $m = 2$ or $m = 1$, as H. Weyl first observed; he called the case $m = 2$ the *limit-circle case,* and the case $m = 1$ the *limit-point case.* We shall explain this nomenclature in §129.

Boundary conditions for the case of an interval with two singular end-points can be constructed without difficulty using the methods of this section, and so we omit this topic.

To conclude the section we look briefly at the question of self-adjoint extensions L_θ which are real relative to the operator of complex conjugation in $L^2(0, \infty)$ (see §50).

Since the operator L^* is real relative to the operator of complex conjugation because the coefficients of the generating differential expression are real, it follows that L_θ will be real if and only if the manifold $D(L_\theta)$ contains the function $\overline{\varphi(t)}$ whenever it contains $\varphi(t)$.

Using this fact and the formula (1) it is easy to show that:

the operator L_θ will be real if and only if the unitary matrix $\theta = [\theta_{ik}]$ is symmetric under transposition, i.e., if and only if $\theta_{ik} = \theta_{ki}$ $(i, k = 1, 2, \ldots, m)$.

This implies, in particular, that, if the deficiency index is $(1, 1)$, then every self-adjoint extension is real.

In §128 wc shall need to use only the fact that there is at least one real extension, a result which follows immediately from the formula (1) if θ is taken to be the unit matrix.

§128. Resolvents of self-adjoint extensions

In this section we shall prove that the resolvents of self-adjoint extensions of a differential operator L are integral operators, and we shall indicate the classes to which the kernels of these integral operators belong when the deficiency index of the operator L takes various values. We begin with two auxiliary propositions.

Lemma 1. *Suppose that the resolvent of some self-adjoint operator which operates in $L^2(0, \infty)$ and is real relative to the operator of complex conjugation has the form, on all functions $g(t) \in L^2(0, \infty)$ which have* *compact support,*

$$R(\lambda)g = \int_0^\infty K(t, s; \lambda)g(s)\, ds,$$

where the kernel $K(t, s; \lambda)$ is continuous with respect to both s and t when $s>0$ and $t>0$ and $t \neq s$.

Then

$$K(t, s; \lambda) = K(s, t; \lambda).$$

Proof. This follows immediately from the relation $R^*(\lambda) = JR(\lambda)J$ (see §50) and the fact that the function $g(t)$ can be chosen arbitrarily.

Lemma 2. Every homogeneous and additive (but, in general, unbounded) functional **L** *which is defined on all functions $g(t) \in L^2(0, \infty)$ with compact support and which is bounded in $L^2(0, a)$ for every finite $a>0$ admits the representation*

$$\mathbf{L}(g) = \int_0^\infty g(s)\overline{h(s)}\, ds, \tag{1}$$

where $h(t)$ is a function which is uniquely determined by the functional **L** *and which belongs to $L^2(0, a)$ for any finite $a>0$.*

Proof. This follows from Riesz's theorem on the general form of a linear functional (see §19), according to which **L** on functions $g(t) \in L^2(0, a)$ has the representation

$$\mathbf{L}(g) = \int_0^a g(s)\overline{h_a(s)}\, ds \qquad (h_a(s) \in L^2(0, a)). \tag{2}$$

The function $h_a(t)$ appearing in this representation is 'prolonged', when a increases, in the sense that the inequality $a_1 < a_2$ implies the equality

$$h_{a_1}(t) = h_{a_2}(t)$$

for all $t < a_1$ (except perhaps for a set of measure zero). This is because, for any function $g(t) \in L^2(0, \infty)$ the two representations

$$\mathbf{L}(g) = \int_0^{a_1} g(s)\overline{h_{a_1}(s)}\, ds$$

and

$$\mathbf{L}(g) = \int_0^{a_2} g(s)\overline{h_{a_2}(s)}\, ds = \int_0^{a_1} g(s)\overline{h_{a_1}(s)}\, ds$$

hold simultaneously, and so

$$\int_0^{a_i} \{h_{a_2}(s) - h_{a_1}(s)\}\overline{g(s)}\, ds = 0.$$

This property of the function h_a enables the formula (2) to be written in the form (1). Q.E.D.

Theorem 1. The resolvent $\tilde{R}(\lambda)$ of any self-adjoint extension $\tilde{L}$ of a differential operator L with deficiency index (m, m) at regular points λ is an integral operator.

Proof. We carry out the proof first for some real, self-adjoint extension, and then show that if the theorem holds for one self-adjoint extension, then it is true for all self-adjoint extensions.

Let $\tilde{L}_0$ be a real, self-adjoint extension of the operator L, and let $\tilde{R}^0(\lambda)$ be the corresponding resolvent (λ being a fixed, regular point of the operator $\tilde{L}$).

First of all we note that the element $\tilde{R}^0(\lambda)g \quad (g(t) \in L^2(0, \infty))$ has to be sought in the manifold of solutions of the inhomogeneous equation

$$l[y] - \lambda y = g. \tag{3}$$

We choose a fundamental system of solutions

$$u_i(t; \lambda) \qquad (i = 1, 2, \ldots, 2n)$$

of the corresponding homogeneous equation

$$l[u] - \lambda u = 0 \tag{4}$$

so that the m solutions $u_1(t; \lambda), u_2(t; \lambda), \ldots, u_m(t; \lambda)$ belong to $L^2(0, \infty)$ —— the solutions $u_{m+1}(t; \lambda), \ldots, u_{2m}(t; \lambda)$ and all non-trivial linear combinations of them will then not belong to $L^2(0, \infty)$; here we do not presuppose, as we did in §127, that the solutions

$$u_i(t; \lambda) \qquad (i = 1, 2, \ldots, m)$$

are orthogonal.

By §123, the general solution of equation (3) is of the form

$$\varphi(t) = \sum_{k=1}^{2n} u_k(t; \lambda) \int_0^t v_k(s; \lambda) g(s)\, ds + \sum_{k=1}^{2n} c_k u_k(t; \lambda), \tag{5}$$

where the $v_k(t; \lambda) \quad (k = 1, 2, \ldots, 2n)$ form a fundamental system of solutions of equation (4) which are adjoint to the system

$$u_k(t; \lambda) \qquad (k = 1, 2, \ldots, 2n).$$

In formula (5) the constants c_k $(k=1,2,\ldots,2n)$ are uniquely determined once the condition that $\varphi(t)=\tilde{R}^0(\lambda)g$ is imposed. We now deal with the determination of the values of the constants for which this condition is satisfied, on the assumption that $g(t)$ is a function with compact support.

First of all we must ensure that $\varphi(t)$ belongs to $L^2(0,\infty)$. But if $g(t)=0$ for $t\geqslant a$, then, for sufficiently large t,

$$\varphi(t)=\sum_{k=1}^{2n} u_k(t;\lambda)\int_0^a v_k(s;\lambda)g(s)\,ds+\sum_{k=1}^{2n} c_k u_k(t;\lambda),$$

and $\varphi(t)$ will belong to $L^2(0,\infty)$ if and only if the $2n-m$ constants c_k $(k=m+1,m+2,\ldots,2n)$ are chosen in accordance with the formulae

$$c_k=-\int_0^a v_k(s;\lambda)g(s)\,ds=-\int_0^\infty v_k(s;\lambda)g(s)\,ds$$
$$(k=m+1,m+2,\ldots,2n).$$

Having chosen the constants in this way, we obtain for $\tilde{R}^0(\lambda)$ the representation

$$\begin{aligned}\tilde{R}^0(\lambda)g&=\sum_{k=1}^{2n} u_k(t;\lambda)\int_0^t v_k(s;\lambda)g(s)\,ds\\&\quad+\sum_{k=1}^{m} c_k u_k(t;\lambda)-\sum_{k=m+1}^{2n} u_k(t;\lambda)\int_0^\infty v_k(s;\lambda)g(s)\,ds\\&=\sum_{k=1}^{m} u_k(t;\lambda)\int_0^t v_k(s;\lambda)g(s)\,ds\\&\quad-\sum_{k=m+1}^{2n} u_k(t;\lambda)\int_t^\infty v_k(s;\lambda)g(s)\,ds+\sum_{k=1}^{m} c_k u_k(t;\lambda),\end{aligned}$$

or

$$\tilde{R}^0(\lambda)g=\int_0^\infty K(t,s;\lambda)g(s)\,ds+\sum_{k=1}^{m} c_k u_k(t;\lambda), \tag{6}$$

where

$$K(t,s;\lambda)=\begin{cases}\displaystyle\sum_{k=1}^{m} u_k(t;\lambda)v_k(s;\lambda) & (s\leqslant t)\\ \displaystyle-\sum_{k=m+1}^{2n} u_k(t;\lambda)v_k(s;\lambda) & (s>t).\end{cases}$$

In the representation (6) $g(t)\in L^2(0,\infty)$ is any function which is equal to 0 outside some finite interval, and the c_k $(k=1,2,\ldots,m)$ are the constants which correspond to this function. It follows immediately from this representation that these constants c_k are homogeneous and additive functionals defined on all functions $g \in L^2(0, \infty)$ with compact support. We shall prove that these functionals $c_k \quad (k = 1, 2, \ldots, m)$ satisfy the conditions of Lemma 2, *i.e.*, they are bounded in $L^2(0, a)$ for every $a < \infty$. To do this, we apply to both sides of the equality (6) the operator P_a of projection on to the sub-space $L^2(0, a)$, and then multiply scalarly by $u_i(t; \lambda) \quad (i = 1, 2, \ldots, m)$ to obtain

$$(P_a\tilde{R}^0(\lambda)g, u_i) = (K_a g, u_i) + \sum_{k=1}^{m} c_k(P_a u_k, u_i) \qquad (i = 1, 2, \ldots, m), \tag{7}$$

where

$$K_a g = P_a \int_0^a K(t, s; \lambda)g(s)\, ds.$$

In the system of m equations (7) in the m unknowns c_k $(k = 1, 2, \ldots, m)$ the norm of the operator $P_a\tilde{R}^0(\lambda)$ is bounded for all positive $a < \infty$ (by a number which does not depend on a), and the norm of the integral operator K_a in $L^2(0, a)$ with a Hilbert–Schmidt kernel is bounded for every $a < \infty$ (by a number which does depend on a), and finally the Gram determinant

$$\det[(P_a u_k, u_i)]_{i,k=1}^{\infty}$$

is not equal to 0 because the solutions $u_k(t; \lambda) \quad (k = 1, 2, \ldots, m)$ are linearly independent. Hence it follows that the constants c_k $(k = 1, 2, \ldots, m)$ are bounded functionals in $L^2(0, a)$ for every $a < \infty$, and therefore, by Lemma 2, they can be expressed in the form

$$c_k(g) = \int_0^\infty g(s)\psi_k(s; \lambda)\, ds \qquad (k = 1, 2, \ldots, m),$$

where the functions $\psi_k(s; \lambda) \quad (k = 1, 2, \ldots, m)$ belong to $L^2(0, a)$ for any $a < \infty$. The representation (6) now takes the form

$$\tilde{R}^0(\lambda)g = \int_0^\infty K(t, s; \lambda)g(s)\, ds + \sum_{k=1}^{m} u_k(t; \lambda) \int_0^\infty g(s)\psi_k(s; \lambda)\, ds,$$

or

$$\tilde{R}^0(\lambda)g = \int_0^\infty K_0(t, s; \lambda)g(s)\, ds, \tag{8}$$

where

$$K_0(t, s; \lambda) = \begin{cases} \sum_{k=1}^{m} u_k(t; \lambda)\{v_k(s; \lambda) + \psi_k(s; \lambda)\} & (s \leq t), \\ \sum_{k=1}^{m} u_k(t; \lambda)\psi_k(s; \lambda) - \sum_{k=m+1}^{2n} u_k(t; \lambda)v_k(s; \lambda) & (s > t). \end{cases} \tag{9}$$

Formula (8) establishes that the resolvent $\tilde{R}^0(\lambda)$ is an integral operator on all functions $g(t) \in L^2(0, \infty)$ which have compact support.

It does not follow from this fact alone that $\tilde{R}^0(\lambda)$ is an integral operator. From the formula (8) it follows only that, because the operator $\tilde{R}^0(\lambda)$ is bounded,

$$\tilde{R}^0(\lambda)g = \underset{n\to\infty}{\text{l.i.m.}} \int_0^n K_0(t, s; \lambda)g(s)\, ds \tag{10}$$

for any function $g \in L^2(0, \infty)$.

In order to obtain from this the desired representation for $\tilde{R}^0(\lambda)g$ in the form

$$\tilde{R}^0(\lambda)g = \int_0^\infty K_0(t, s; \lambda)g(s)\, ds \qquad (g(t) \in L^2(0, \infty)), \tag{11}$$

it suffices to show that the integral

$$\int_0^\infty K_0(t, s; \lambda)g(s)\, ds$$

exists for any function $g(t) \in L^2(0, \infty)$, and to do this it suffices, in turn, to show that the integral

$$\int_0^\infty |K_0(t, s; \lambda)|^2\, ds \qquad (0 \le t < \infty) \tag{12}$$

exists, *i.e.*, that the kernel $K(t, s; \lambda)$ regarded as a function of s belongs to $L^2(0, \infty)$ for every t.

Returning to the formula (9), we observe that the kernel $K_0(t, s; \lambda)$ regarded as a function of t belongs to $L^2(0, \infty)$ for every s (because $u_k(t; \lambda) \in L^2(0, \infty)$ for $k = 1, 2, \ldots, m$); but since by Lemma 1 the function $K_0(t, s; \lambda)$ corresponding to the resolvent of a real symmetric extension is symmetric with respect to the arguments t and s, the existence of the integral (12) is established and the representation (11) has been proved.

The theorem is proved in a similar way for the resolvent $\tilde{R}(\lambda)$ of any self-adjoint (but not necessarily real) extension $\tilde{L}$ for any real, regular value of λ, if we use† instead of the equality

$$K_0(s, t; \lambda) = K_0(t, s; \lambda)$$

†An important point in the above argument is that the transition from (10) to (11) is effected by relying on the symmetry of the kernel $K_0(t, s; \lambda)$, and thanks to this the question of the behaviour of the functions $v_k(s; \lambda)$ $(k = m+1, m+2, \ldots, 2n)$ and $\psi_k(s; \lambda)$ $(k = 1, 2, \ldots, m)$ at $s = \infty$ need not be discussed. Moreover, the symmetry of the kernel $K_0(t, s; \lambda)$ enables us to establish various properties of these functions at infinity.

the relation

$$\tilde{K}(s, t; \lambda) = \overline{\tilde{K}(t, s; \lambda)},$$

where $\tilde{K}(s, t; \lambda)$ is the kernel of the resolvent $\tilde{R}(\lambda)$.

To complete the proof of the theorem it remains to obtain a representation for $\tilde{R}(\lambda)$ when $\mathscr{I}(\lambda) \neq 0$.

Since an element $\tilde{R}(\lambda)g \quad (g(t) \in L^2(0, \infty))$ belongs to the manifold of solutions of equation (3), it can be expressed in the form

$$\tilde{R}(\lambda)g = \tilde{R}^0(\lambda)g + \sum_{k=1}^{2n} c_k u_k, \tag{13}$$

where $\tilde{R}^0(\lambda)$ is the resolvent of some real, self-adjoint extension. From this representation corresponding to the choice of the fundamental system $u_k(t; \lambda) \quad (k = 1, 2, \ldots, 2n)$ it follows that, when $k > m$, the coefficients $c_k = 0$, and when $k \leq m$, they are linear (*i.e.*, homogeneous, additive, and bounded) functionals in $L^2(0, \infty)$, and therefore they admit the representation

$$c_k = c_k(g) = \int_0^\infty g(s)\chi_k(s; \lambda)\, ds \qquad (k = 1, 2, \ldots, m), \tag{14}$$

where the functions $\chi_k(s; \lambda) \quad (k = 1, 2, \ldots, m)$ belong to $L^2(0, \infty)$.

On substituting (14) into (13) and introducing the kernel $\tilde{K}(t, s; \lambda)$ by the equality

$$\tilde{K}(t, s; \lambda) = K_0(t, s; \lambda) + \sum_{k=1}^{m} u_k(t; \lambda)\chi_k(s; \lambda), \tag{15}$$

we finally obtain an integral representation of the resolvent in the form

$$\tilde{R}(\lambda)g = \int_0^\infty \tilde{K}(t, s; \lambda)g(s)\, ds \qquad (g(t) \in L^2(0, \infty)),$$

and Theorem 1 has been proved completely.†

Theorem 2. The kernel $\tilde{K}(t, s; \lambda)$ of the resolvent $\tilde{R}(\lambda)$ of any self-adjoint extension $\tilde{L}$ of an operator L with deficiency index (m, m) satisfies, at regular points, the conditions

$$\int_0^\infty |\tilde{K}(t, s; \lambda)|^2\, ds < \infty, \qquad \int_0^\infty |\tilde{K}(t, s; \lambda)|^2\, dt < \infty, \tag{16}$$

and if the operator L is quasi-regular, i.e., if $m = 2n$, it satisfies the condition

$$\int_0^\infty \int_0^\infty |\tilde{K}(t, s; \lambda)|^2\, ds\, dt < \infty. \tag{17}$$

† A generalization of this theorem to the case of non-orthogonal resolvents has been obtained by A. V. Shtraus [115].

Proof. The property (16) follows immediately from the formula (15) and the fact that the kernel $K_0(t, s; \lambda)$ belongs to $L^2(0, \infty)$ both as a function of t and as a function of s, and that the functions $\chi_k(s; \lambda)$ $(k = 1, 2, \ldots, m)$ belong to $L^2(0, \infty)$.

Turning to the proof of property (17), we first show that, for any deficiency index (m, m) $(n \leqslant m \leqslant 2m)$ of the operator L, the functions

$$\psi_k(t; \lambda) \qquad (k = 1, 2, \ldots, m)$$

and

$$v_k(t; \lambda) \qquad (k = m+1, m+2, \ldots, 2n),$$

which appear in the formula (9) for $K_0(t, s; \lambda)$ belong to the space $L^2(0, \infty)$. To do this, we again make use of the symmetry property of the kernel $K_0(t, s; \lambda)$. This property implies that, for $s \leqslant t$,

$$\sum_{k=1}^{m} u_k(s; \lambda)\psi_k(t; \lambda) - \sum_{k=m+1}^{2n} u_k(s; \lambda)v_k(t; \lambda)$$
$$= \sum_{k=1}^{m} \{v_k(s; \lambda) + \psi_k(s; \lambda)\}u_k(t; \lambda). \tag{18}$$

We choose numbers $s_1, s_2, \ldots, s_{2n}$ such that

$$\det\,[u_k(s_i; \lambda)]_{i,k=1}^{2n} \neq 0$$

(this is possible because the functions $u_k(s; \lambda)$ are linearly independent), and we put $s = s_i$ $(i = 1, 2, \ldots, 2n)$ in the equality (18).

By solving this system of $2n$ equations for the $2n$ functions

$$\psi_k(t; \lambda) \qquad (k = 1, 2, \ldots, m)$$

and

$$v_k(t; \lambda) \qquad (k = m+1, m+2, \ldots, 2n),$$

we find that each of these functions is some linear combination of the functions $u_1(t; \lambda), \ldots, u_m(t; \lambda)$, and consequently belongs to the space $L^2(0, \infty)$.

Now we can establish the property (17). In the case of a quasi-regular operator $(m = 2n)$, all the terms on the right in (9) belong to the space L^2 of functions of two variables in the first quadrant $t > 0$, $s > 0$, and so

$$\int_{-\infty}^{\infty}\int_{-\infty}^{\infty} |K_0(t, s; \lambda)|^2 \, ds \, dt < \infty. \tag{19}$$

(15) and (19) then imply the property (17). Q.E.D.

It follows from this theorem that, in the case of a quasi-regular operator, the resolvent of any self-adjoint extension $\tilde{L}$ of the operator L

is determined by a Hilbert–Schmidt kernel, and so it is a completely continuous operator.

Theorem 2 also implies that the kernels of the real and imaginary parts of the resolvent of any self-adjoint extension of the operator at regular points λ are Carleman kernels.

We further observe that the argument used in proving Theorem 2 also yields the following result.

Theorem 3. *Let m be the maximum number of linearly independent solutions of the equation* (4): $l[u]-\lambda u=0$ *which belong to* $L^2(0,\infty)$. *If the first m functions in a fundamental system of solutions of this equation belong to* $L^2(0,\infty)$, *then at least the last* $2n-m$ *functions in the adjoint fundamental system of solutions of this equation also belong to* $L^2(0,\infty)$.

The next theorem is about a case where the deficiency index of the operator L can be inferred from the number of solutions of the equation (4) which belong to $L^2(0,\infty)$ for an arbitrary value of λ and, in particular, for a real λ.

Theorem 4. *For the operator L of order* $2n$ *to have the deficiency index* $(2n, 2n)$ *it is necessary that the equation*

$$l[u]-\lambda u=0$$

shall have $2n$ *solutions belonging to* $L^2(0,\infty)$ *for any* λ, *and it is sufficient that this equation shall have* $2n$ *solutions belonging to* $L^2(0,\infty)$ *for at least one value of* λ.

Proof. To prove the necessity of the stated condition, we recall that in the case of a quasi-regular operator, the resolvent of a self-adjoint extension $\tilde{L}$ of that operator is a completely continuous operator, and consequently the spectrum of the operator $\tilde{L}$ consists only of eigenvalues λ_r $(r=1,2,3,\ldots)$ with a unique limit-point $\lambda=\infty$. Hence it follows for the operator L that all points of the plane, except possibly the points λ_r $(r=1,2,3,\ldots)$, are points of regular type. In fact, the points λ_r $(r=1,2,3,\ldots)$ are also points of regular type, for otherwise λ_r would be an eigenvalue of the operator L; but this is impossible, since the equation

$$L\varphi-\lambda_r\varphi=0$$

is equivalent to the equation

$$l[\varphi]-\lambda_r\varphi=0$$

under the conditions

$$\varphi(0)=\varphi^{[1]}(0)=\cdots=\varphi^{[2n-1]}(0)=0,$$

and so, by the existence and uniqueness theorem in §123, it has only the solution $\varphi(t)\equiv 0$.

Thus, when the deficiency index is $(2n, 2n)$, the whole λ-plane is a field of regularity for the operator L, and by the theorem on the invariance of the deficiency index (§100), all solutions of the equation (4) for *any* value of λ (non-real or real) belong to $L^2(0,\infty)$.

The sufficiency of the condition stated follows immediately from Theorem 4 in §105. Q.E.D.

It is easy to translate the results so far obtained in this section to the case of an interval with two singular end-points. We outline the proof of the main theorem, Theorem 1, for this case. Here, as in the case of one singular end-point, it suffices to consider the resolvent of some real, self-adjoint extension. Let (see the proof of Theorem 3 in §126) the deficiency numbers of the operators L, $L^{(-)}$, $L^{(+)}$ be m, $m^{(-)}$, $m^{(+)}$ respectively. By Theorem 3 of §126,

$$m^{(-)}=n+p, \qquad m^{(+)}=n+m-p \qquad (0\leqslant p\leqslant n).$$

Choose a fundamental system of solutions of equation (4) such that the first m solutions $u_k(t;\lambda)$ belong to $L^2(-\infty,\infty)$. In the $(2n-m)$-dimensional linear envelope of the remaining solutions of this system there are $n+p-m$ linearly independent solutions which belong to $L^2(-\infty,0)$ and which we denote by $u_{m+1}(t;\lambda),\ldots,u_{n+p}(t;\lambda)$, and $n-p$ linearly independent solutions which belong to $L^2(0,\infty)$ and which we denote by $u_{n+p+1}(t;\lambda),\ldots,u_{2n}(t;\lambda)$. We observe that the functions

$$u_{m+1}(t;\lambda),\ldots,u_{n+p}(t;\lambda),\, u_{n+p+1}(t;\lambda),\ldots,u_{2n}(t;\lambda)$$

are linearly independent, for otherwise some linear combination of them would belong to $L^2(-\infty,\infty)$, from which it would follow that the deficiency number of the operator L was greater than m, contrary to hypothesis.

By means of this chosen fundamental system

$$u_k(t;\lambda) \qquad (k=1,2,\ldots,2n),$$

any solution of the equation

$$l[y]-\lambda y=g,$$

where $g(t)$ is a function of compact support, can be expressed in the form

$$\varphi(t)=\sum_{k=1}^{2n} u_k(t;\lambda)\int_{-\infty}^{t} v_k(s;\lambda)g(s)\,ds+\sum_{k=1}^{2n} c_k u_k(t;\lambda),$$

and if the constants c_k are suitably chosen, this function coincides with $\tilde{R}^0(\lambda)g$.

By an argument similar to that used in the case of one singular end-point we obtain the complete proof of Theorem 1 for the case of two singular end-points.

We now give a simple proposition which sometimes enables the deficiency index of an operator generated by the differential expression

$$-y''+q(x)y \tag{20}$$

to be determined.

Theorem 5. *If $q(x)\geqslant q_0>-\infty$, then the deficiency index of the minimal operator, with one singular end-point, generated by the differential expression* (20) *in* $L^2(0,\infty)$ *is* (1, 1), *and the minimal operator generated by the same differential expression in* $L^2(-\infty,\infty)$ *has the deficiency index* (0, 0), *i.e., it is self-adjoint.*

Proof. We shall suppose that $q(x)\geqslant 0$. Then, by applying the method of successive approximations to the equation

$$u''=q(t)u \qquad (0\leqslant t<\infty),$$

we find that the solution $u(t)$ such that $u(0)=1$, $u'(0)=0$, satisfies the inequality $u(t)\geqslant 1$ for all $t\geqslant 0$, and so it does not belong to $L^2(0,\infty)$.

Thus the equation

$$-u''+q(t)u-\lambda u=0,$$

when $\lambda=0$, has a solution which does not belong to $L^2(0,\infty)$, and so, by Theorem 4 of this section, the operator with minimal domain which is generated by the differential expression (20) on the positive half-line has the deficiency index (1, 1). The operator with minimal domain which is generated by the differential expression (20) on the negative half-line will evidently have the same deficiency index (1, 1). So, applying Theorem 3 of §126, we obtain the result that the deficiency index of the operator with minimal domain generated by the differential expression (20) on the whole line is (0, 0).

Clearly, we would have arrived at this same result by noting that the requirement that the function $q(x)$ be non-negative is equivalent to requiring that $q(x)$ be semi-bounded below.

§129. Inversion formulae associated with second-order differential operators

In §83 it was shown that every self-adjoint operator A with a simple spectrum generates an isometric mapping V of the space H on to some space L^2_σ. Under this mapping, the operator A turns into the operator of multiplication by the independent variable.

If E_λ is the resolution of the identity and if g is some generating element for the operator A, then $\sigma(\lambda)=(E_\lambda g, g)$, and the isometric correspondence referred to is defined by the formulae

$$\Phi(\lambda)=Vf, \tag{1}$$

$$f=V^{-1}\Phi(\lambda)=\int_{-\infty}^{\infty}\Phi(\lambda)\,dE_\lambda g \tag{2}$$

where the element f and the function Φ traverse the spaces H and L^2_σ respectively.

If also $f\in D(A)$, then, as well as (2), the formula

$$Af=\int_{-\infty}^{\infty}\lambda\Phi(\lambda)\,dE_\lambda g$$

holds.

We shall call (1) and (2) the *inversion formulae* associated with a self-adjoint operator A.

In paragraph C of §89 we obtained as the inversion formulae associated with the operator of differentiation on a real line, the mutually inverse Fourier–Plancherel transformations of the space $L^2(-\infty, \infty)$ on to itself:

$$\Phi(\lambda)=\underset{N\to\infty}{\text{l.i.m.}}\frac{1}{\sqrt{(2\pi)}}\int_{-N}^{N}f(t)e^{i\lambda t}\,dt,$$

$$f(t)=\underset{N\to\infty}{\text{l.i.m.}}\frac{1}{\sqrt{(2\pi)}}\int_{-N}^{N}\Phi(\lambda)e^{-i\lambda t}\,d\lambda.$$

In this case a certain improper element was taken as the generating element.

An operator with a multiple spectrum also generates inversion formulae similar to (1) and (2).

The problem in this section and the next is to obtain the inversion formulae associated with singular differential operators on the half-line. For second-order operators this problem was first solved by H. Weyl.†

† H. Weyl [126]. Weyl's results were also obtained by B. M. Levitan [70] using another method.

We too shall begin by examining second-order operators, but we shall use M. G. Krein's method of direction functionals.† This method enables us, in particular, to obtain inversion formulae for differential operators of arbitrary order (see §130).

Let, then, L be the differential operator with minimal domain which is generated by the differential expression

$$l = -\frac{d}{dt}\left(p\frac{d}{dt}\right) + q$$

on the interval $(0, \infty)$ with the one singular end-point.

We shall assume that the deficiency index of the operator L is $(1, 1)$. We shall see that in this case the spectrum of every self-adjoint extension of this operator is simple.‡

Let $\tilde{L}$ be a self-adjoint extension of the operator L, defined by the boundary conditions††

$$[p(t)\varphi'(t)]_{t=0} = \theta\varphi(0) \qquad (\mathcal{I}(\theta) = 0)$$

and let $u_1(t; \lambda)$ and $u_2(t; \lambda)$ be the solutions of the equation

$$l[u] - \lambda u = 0 \qquad (\mathcal{I}(\lambda) = 0)$$

which satisfy the initial conditions

$$u_1(0; \lambda) = 1, \qquad [p(t)u_1'(t; \lambda)]_{t=0} = 0,$$
$$u_2(0; \lambda) = 0, \qquad [p(t)u_2'(t; \lambda)]_{t=0} = 1.$$

Let $\mathcal{D}$ denote the set of all functions $f(t) \in L^2(0, \infty)$ which have compact support; we define in $\mathcal{D}$ the homogeneous and additive functional $\Phi(\cdot; \lambda)$ by

$$\Phi(f; \lambda) = \int_0^\infty f(t)u(t; \lambda)\, dt,$$

where

$$u(t; \lambda) = u_1(t; \lambda) + \theta u_2(t; \lambda).$$

† See M. G. Krein [59, 61]. In this connection see also M. S. Livshits [74] and A. Ya Povzner [101]. On the method of direction functionals applied to operators with a spectrum of infinite multiplicity, see Yu. M. Berezansky [13] and also F. S. Rofe-Beketov [106].

‡ If the deficiency index of the operator L is $(2, 2)$, then every self-adjoint extension of this operator has a discrete spectrum. In this case the question of the structure of the inversion formulae does not arise, since the rôle of the formula (2) is played by the expansion of the function as a series of orthogonal eigenfunctions, and the rôle of the formula (1) by the expression for the Fourier coefficients in this expansion.

†† Thus we are leaving out just one extension, *viz.*, that determined by the boundary condition $\varphi(0) = 0$ (corresponding to $\theta = \infty$). The inversion formulae obtained below can, by a limiting process, be extended to cover this case too.

Following M. G. Krein we shall call $\Phi(f;\lambda)$ the *direction functional.*

The direction functional has the following three properties, which are essential for our purpose.

1°. $\Phi(f;\lambda)$ is an analytic function of λ $(-\infty<\lambda<\infty)$ for any function $f\in\mathcal{D}$.

2°. If the equality $\Phi(f;\lambda)=0$ holds for some function $f\in\mathcal{D}$ and for some real value of λ, then the equation

$$(\tilde{L}-\lambda I)\varphi=f,$$

will have a solution in the class of functions which have compact support.

3°. If $f\in D(\tilde{L})$ is a function of compact support, then, for any real λ,

$$\Phi(\tilde{L}f;\lambda)=\lambda\Phi(f;\lambda).$$

Only properties 2° and 3° are in need of proof. We prove 2° first. Let $f(t)=0$ for $t>a$ and let

$$\Phi(f;\lambda_0)=0. \tag{3}$$

Let $\varphi(t)$ $(0\leqslant t<\infty)$ denote the solution of the equation

$$l[\varphi]-\lambda_0\varphi=0$$

which satisfies the conditions

$$\varphi(a)=p(a)\varphi'(a)=0;$$

then by the uniqueness theorem, $\varphi(t)=0$ for $t\geqslant a$.

Using Lagrange's identity, we obtain

$$\Phi(f;\lambda_0)=\int_0^\infty\{l[\varphi]-\lambda_0\varphi\}u(t;\lambda_0)\,dt=[\varphi,\bar{u}]_0^\infty$$

$$=-[\varphi,\bar{u}]_0=\left[p\left(u\frac{d\varphi}{dt}-\varphi\frac{du}{dt}\right)\right]_{t=0}=[p(t)\varphi'(t)-\theta\varphi(t)]_{t=0}$$

which, combined with (3), gives

$$[p(t)\varphi'(t)]_{t=0}=\theta\varphi(0),$$

and so $\varphi(t)\in D(\tilde{L})$, and property 2° has been established.

Property 3° follows from the equality

$$\Phi(\tilde{L}f;\lambda)=\int_0^\infty l[f]u(t;\lambda)\,dt=[f,\bar{u}]_0^\infty+\lambda\int_0^\infty f(t)u(t;\lambda)\,dt=\lambda\Phi(f;\lambda).$$

The direction functional $\Phi(f;\lambda)$ generates a mapping of the linear manifold $\mathcal{D}$ of the functions $f(t)$ on to some linear manifold $\mathcal{D}'$ of functions $\Phi(\lambda)=\Phi(f;\lambda)$. If $f(t)$ also belongs to $D(\tilde{L})$, then, by property 3° the function corresponding to $\tilde{L}f$ under this mapping will be $\lambda\Phi(\lambda)$.

This circumstance allows us to presuppose that the spectrum of the operator $\tilde{L}$ is simple, and that, with a suitable choice of the generating function $g = g(t)$, the mapping effected by the operator V of formula (1) is an extension of the mapping effected by the functional $\Phi(f; \lambda)$. This supposition implies that, with a suitable choice of the generating function, the first of the inversion formulae associated with the operator $\tilde{L}$ will have the form

$$\Phi(\lambda) = \int_0^\infty f(t)u(t; \lambda)\, dt \qquad (f(t) \in \mathcal{D}).$$

This formula is analogous to the Fourier–Plancherel formula.

We now leave heuristic considerations aside and set about solving our problem. We shall need two results due to M. G. Krein (see [61]).

Lemma. *For any finite interval Δ_0 of the λ-axis, there is a function $\psi(t) \in \mathcal{D}$ such that the function $\Phi(\lambda) = \Phi(\psi; \lambda)$ is never equal to 0 in the interval Δ_0.*

Proof. Let λ_0 be a fixed point of the interval Δ_0, and let $\chi_0(t) \in \mathcal{D}$ be a function such that

$$\Phi(\chi_0; \lambda_0) \neq 0.$$

If $\Phi(\chi_0; \lambda_0) \neq 0$ throughout the whole interval Δ_0, then it suffices to take $\psi = \chi_0$ to prove this lemma. But if, at some point $\lambda_1 \in \Delta_0$,

$$\Phi(\chi_0; \lambda_1) = 0,$$

then, by property 2° there is a function $\chi_1 \in D(\tilde{L})$ which is of compact support and is such that

$$\tilde{L}\chi_1 - \lambda_1\chi_1 = \chi_0.$$

By applying the direction functional to both members of this equality, and using property 3°, we obtain

$$\lambda\Phi(\chi_1; \lambda) - \lambda_1\Phi(\chi_1; \lambda) = \Phi(\chi_0; \lambda),$$

and therefore

$$\Phi(\chi_1; \lambda) = \frac{\Phi(\chi_0; \lambda)}{\lambda - \lambda_1}.$$

Thus, replacing the function χ_0 by the function χ_1, we see that the function $\Phi(\chi_1; \lambda)$ is not equal to 0 at $\lambda = \lambda_0$, and at $\lambda = \lambda_1$ it has a root of multiplicity one less than that of the root of $\Phi(\chi_0; \lambda)$ at this point. Since the analytic function $\Phi(\chi_0; \lambda)$ has only a finite number of zeros in the finite interval Δ_0 and each of them is of finite multiplicity, so, continuing this process if necessary, we arrive at a function $\psi \in \mathcal{D}$ such that the functional $\Phi(\psi; \lambda)$ is not equal to 0 at any point $\lambda \in \Delta_0$. Q.E.D.

Theorem 1. *Let Δ_0 be a finite, closed interval, and let $\psi \in \mathcal{D}$ be a function such that*

$$\Phi(\psi; \lambda) \neq 0$$

for any $\lambda \in \Delta_0$. Let E_λ be the resolution of the identity for the operator $\tilde{L}$, and let

$$g(t) = \int_{\Delta_0} \frac{1}{\Phi(\psi; \lambda)} \, dE_\lambda \psi.$$

Then, for any function $f \in \mathcal{D}$ and for any interval $\Delta \subseteq \Delta_0$, the formula

$$E(\Delta) f = \int_\Delta \Phi(f; \lambda) \, dE_\lambda g \tag{4}$$

holds.

Proof. If we introduce instead of the direction functional $\Phi(f; \lambda)$ a functional $F(f; \lambda)$ defined by

$$F(f; \lambda) = \frac{\Phi(f; \lambda)}{\Phi(\psi; \lambda)},$$

then the formula to be proved is equivalent to the equality

$$E(\Delta) f = \int_\Delta F(f; \lambda) \, dE_\lambda \psi, \tag{5}$$

which we proceed to prove.

Suppose that the left-hand end-points of the intervals Δ_0 and Δ coincide; let α be this common point. We shall consider an element $w_\mu = w_\mu(t)$ of the space $L^2(0, \infty)$ defined by

$$w_\mu = \int_\alpha^\mu dE_\lambda f - \int_\alpha^\mu F(f; \lambda) \, dE_\lambda \psi$$

as a function of the parameter μ in the interval Δ_0.

To prove (5) we have to show that the element $w_\mu = 0$ when $\mu \in \Delta_0$, and to do this it clearly suffices to show that w_μ is a continuous function of μ in Δ_0 and then that

$$\frac{d}{d\mu} w_\mu = 0 \tag{6}$$

everywhere in Δ_0. The last equality is to be understood in the sense of strong convergence in $L^2(0, \infty)$, *i.e.*, in the sense that

$$\lim_{\delta\to 0} \frac{1}{\delta^2} \|w_{\mu+\delta} - w_\mu\| = \lim_{\delta\to 0} \frac{1}{\delta^2} \int_0^\infty |w_{\mu+\delta}(t) - w_\mu(t)|^2 \, dt = 0.$$

Since E_λ is left-continuous, the vector function w_μ is left-continuous. To prove that w_μ is right-continuous, we use the identity

$$\Phi(f - F(f; \mu)\psi; \mu) = \Phi(f; \mu) - F(f; \mu)\Phi(\psi; \mu) = 0,$$

which comes from the definition of the functional $F(f; \mu)$. By property 2° it follows from this identity that the element $f - F(f; \mu)\psi$ can be expressed in the form

$$f - F(f; \mu)\psi = \tilde{L}\varphi - \mu\varphi. \tag{7}$$

We now introduce the projection operator

$$E(\Delta_{\mu,\delta}) = E_{\mu+\delta} - E_\mu \qquad (\delta \geqslant 0).$$

$E(\Delta_{\mu,0})\varphi$ is either 0 or it is an eigenvector of the operator $\tilde{L}$ corresponding to the eigenvalue μ. In either case

$$E(\Delta_{\mu,0})(\tilde{L}\varphi - \mu\varphi) = \tilde{L}E(\Delta_{\mu,0})\varphi - \mu E(\Delta_{\mu,0})\varphi = 0,$$

and therefore

$$E(\Delta_{\mu,0})f - F(f; \mu)E(\Delta_{\mu,0})\psi = 0.$$

Hence it follows that w_μ is continuous in the interval Δ_0.

To prove (6) we have to estimate $\|w_{\mu+\delta} - w_\mu\|$. Suppose, for example, that $\delta > 0$; the case $\delta < 0$ is treated similarly. Because w_μ is continuous, we have $w_{\mu+0} = w_\mu$. Therefore

$$\begin{aligned} \|w_{\mu+\delta} - w_\mu\| &= \|w_{\mu+\delta} - w_{\mu+0}\| \\ &= \left\| E(\Delta_{\mu+0,\delta})f - \int_{\mu+0}^{\mu+\delta} F(f; \lambda)\, dE_\lambda \psi \right\| \\ &\leqslant \|E(\Delta_{\mu+0,\delta})[f - F(f; \mu)\psi]\| \\ &\quad + \left\| E(\Delta_{\mu+0,\delta})F(f; \mu)\psi - \int_{\mu+0}^{\mu+\delta} F(f; \lambda)\, dE_\lambda \varphi \right\|. \end{aligned} \tag{8}$$

We consider separately first the first term and then the second term on the right-hand side of this inequality. For the first term we have, by (7), the estimate

$$\begin{aligned} \|E(\Delta_{\mu+0,\delta})[f - F(f; \mu)\psi]\|^2 &= \|E(\Delta_{\mu+0,\delta})[\tilde{L}\varphi - \mu\varphi]\|^2 \\ &= \int_{\mu+0}^{\mu+\delta} |\lambda - \mu|^2 \, d(E_\lambda\varphi, \varphi) \leqslant \delta^2 \|E(\Delta_{\mu+0,\delta})\varphi\|^2. \end{aligned}$$

Therefore

$$\lim_{\delta \searrow 0} \frac{1}{\delta} \|E(\Delta_{\mu+0,\delta})[f - F(f;\mu)\psi]\| = 0. \tag{9}$$

Turning now to the second term of the right-hand side in the inequality (8), we obtain

$$\left\| E(\Delta_{\mu+0,\delta})F(f;\mu)\psi - \int_{\mu+0}^{\mu+\delta} F(f;\lambda)\, dE_\lambda \psi \right\|^2$$

$$= \left\| \int_{\mu+0}^{\mu+\delta} \{F(f;\mu) - F(f;\lambda)\}\, dE_\lambda \psi \right\|^2$$

$$\leqslant \int_{\mu+0}^{\mu+\delta} |F(f,\mu) - F(f;\lambda)|^2\, d(E_\lambda \psi, \psi) \leqslant M^2 \delta^2 (E(\Delta_{\mu+0,\delta})\psi, \psi), \tag{10}$$

where M is the maximum absolute value of the derivative $\partial F(f;\lambda)/\partial\lambda$ in the interval Δ_0.

From (10) it follows that

$$\lim_{\delta \searrow 0} \frac{1}{\delta} \left\| E(\Delta_{\mu+0,\delta})F(f;\mu)\psi - \int_{\mu+0}^{\mu+\delta} F(f;\lambda)\, dE_\lambda \psi \right\| = 0. \tag{11}$$

(8), (9), (11) imply (6), which in turn implies that $w_\mu = 0$, from which the equality (5) follows, as required. Q.E.D.

We can now turn to the main theorem of this section.

Theorem 2. *Let L, $\tilde{L}$, and $u(t;\lambda)$ have the meanings given in the beginning of this section. Then the inversion formulae associated with the operator $\tilde{L}$ have the form*

$$\Phi(\lambda) = \int_0^\infty f(t) u(t;\lambda)\, dt \tag{12}$$

$$f(t) = \int_{-\infty}^{\infty} \Phi(\lambda) u(t;\lambda)\, d\sigma(\lambda), \tag{13}$$

where $\sigma(\lambda)$ is a non-decreasing function which is determined by the operator $\tilde{L}$ (and is called the *spectral function* of this operator), *and $\Phi(\lambda)$ and $f(t)$ traverse the spaces $L^2_\sigma(-\infty, \infty)$ and $L^2(0, \infty)$ respectively.*

Further, if one of the functions $\Phi(\lambda)$, $f(t)$ is equal to zero outside a finite interval, then the integral on the right-hand side of the corresponding formula should be understood in the ordinary sense, but for arbitrary functions $\Phi(\lambda)$, $f(t)$ these integrals are to be understood as the limits in the *metric of the spaces* L^2_σ *and* $L^2(0, \infty)$ *respectively of the proper integrals*

$$\int_0^C f(t)u(t; \lambda)\, dt \qquad \textit{and} \qquad \int_A^B \Phi(\lambda)u(t; \lambda)\, d\sigma(\lambda).$$

Proof. First of all we note that, in the formula (4), we may evidently take,† instead of the element g, the element $g_0 = E(\Delta_0)g$.

We now divide the λ-axis into finite intervals Δ_k $(\pm k = 0, 1, 2, \ldots)$, and for each interval Δ_k we choose an element g_k in the same way as we chose the element g_0 for the interval Δ_0. We put

$$g = \sum_{k=-\infty}^{\infty} g_k,$$

and if the orthogonal series on the right diverges, we shall take g to be an improper element‡ (see §89) satisfying the condition

$$E(\Delta)g = \sum_k E(\Delta)g_k$$

for any finite interval Δ, where the series on the right now contains only a finite number of terms.

If f is an arbitrary function in $\mathscr{D}$, and if Δ is an arbitrary, finite interval of the λ-axis, then

$$\Delta = \Delta_m^* + \sum_{k=m+1}^{n-1} \Delta_k + \Delta_n^* \qquad (\Delta_m^* \subseteq \Delta_m;\ \Delta_n^* \subseteq \Delta_n)$$

and

$$E(\Delta)f = E(\Delta_m^*)f + \sum_{k=m+1}^{n-1} E(\Delta_k)f + E(\Delta_n^*)f.$$

So, for any finite interval Δ of the λ-axis, the formula

$$E(\Delta)f = \int_\Delta \Phi(f; \lambda)\, dE_\lambda g$$

holds, and hence, when $\Delta \to (-\infty, \infty)$, we obtain the equality

$$f(t) = \int_{-\infty}^{\infty} \Phi(f; \lambda)\, dE_\lambda g. \tag{14}$$

Since the manifold $\mathscr{D}$ is dense in $L^2(0, \infty)$, formula (14) implies that the operator L has a simple spectrum, and the function $g = g(t)$ is a generating element.

† In the present case g and g_0 actually coincide.

‡ It turns out, in fact, that the element g is always improper; it behaves like the Dirac δ-function in the sense that $(E(\Delta)h, g) = (E(\Delta)h)(0)$ for any $h \in L^2(0, \infty)$ and any finite interval Δ. (This remark is due to F. S. Rofe-Beketov.)

From the theorem in §83 on the canonical representation of a self-adjoint operator with a simple spectrum it follows that there is an isometric mapping given by

$$f(t)=\int_{-\infty}^{\infty} \Phi(\lambda)\, dE_\lambda g$$

of the space L^2_σ, where $\sigma(\lambda)=(E_\lambda g, g)$, on to the space $L^2(0,\infty)$. Under this isometric mapping, the direction functionals $\Phi(f;\lambda)=\Phi(\lambda)$ correspond to functions $f(t)$ in $\mathscr{D}$. The operator $\tilde{L}$ is thus isomorphic to the operator of multiplication by λ in the space L^2_σ, so that, for $f(t)\in D(\tilde{L})$,

$$\tilde{L}f=\int_{-\infty}^{\infty} \lambda\Phi(\lambda)\, dE_\lambda g.$$

We now show how to derive the formulae (12) and (13).
If $f(t)$ is an arbitrary function in $L^2(0,\infty)$, and if

$$f_n(t)=\begin{cases} f(t) & (t<n), \\ 0 & (t\geqslant n), \end{cases}$$

then, by (14),

$$f_n(t)=\int_{-\infty}^{\infty} \Phi(f_n;\lambda)\, dE_\lambda g. \tag{15}$$

We put

$$\Phi_n(\lambda)=\Phi(f_n;\lambda)=\int_0^n f(t)u(t;\lambda)\, dt. \tag{16}$$

Since the sequence of functions $\{f_n(t)\}$ converges in the metric of $L^2(0,\infty)$ to the function $f(t)$, so, because of the isometry between the spaces $L^2(0,\infty)$ and L^2_σ, the sequence $\{\Phi_n(\lambda)\}$ converges in the metric of L^2_σ to some function $\Phi(\lambda)$. Passing to the limit, in the corresponding metrics, in the equalities (15) and (16), we obtain the formulae

$$\left.\begin{aligned} f(t)&=\int_{-\infty}^{\infty} \Phi(\lambda)\, dE_\lambda g, \\ \Phi(\lambda)&=\int_0^{\infty} f(t)u(t;\lambda)\, dt. \end{aligned}\right\} \tag{17}$$

The second of these formulae is the same as (12); so we have only still to prove formula (13). Clearly, it suffices to prove that (13) is the inversion formula which corresponds to (12) when $\Phi(\lambda)$ traverses the manifold $\mathscr{M}$ of those functions in L^2_σ which have compact support.

Suppose, then, that $\Phi(\lambda)$ belongs to $\mathscr{M}$, and let $f(t)\in L^2(0,\infty)$ be the function which corresponds to $\Phi(\lambda)$ according to (17). We introduce the function

$$\eta_t(s) = \begin{cases} 1 & (s \leq t), \\ 0 & (s > t). \end{cases}$$

Then, because $L^2(0, \infty)$ and L^2_σ are isometric, we have

$$\int_0^\infty f(s)\eta_t(s)\, ds = \int_{-\infty}^\infty \Phi(\lambda)\Phi(\eta_t; \lambda)\, d\sigma(\lambda),$$

or

$$\int_0^t f(s)\, ds = \int_{-\infty}^\infty \Phi(\lambda)\left\{\int_0^t u(s; \lambda)\, ds\right\} d\sigma(\lambda).$$

The left-hand side has a derivative $f(t)$ almost everywhere, and the integral on the right with respect to λ is, effectively, between finite limits, because $\Phi(\lambda) \in \mathcal{M}$. Therefore the derivative of the right-hand side with respect to t is equal to

$$\int_{-\infty}^\infty \Phi(\lambda)u(t; \lambda)\, d\sigma(\lambda).$$

This completes the proof.

We now show how the spectral function $\sigma(\lambda)$ can actually be obtained. To do this, we let $u(t; \lambda)$ denote that solution of the equation

$$l[y] - \lambda y = 0$$

which at $t = 0$ satisfies the boundary condition which determines that particular extension $\tilde{L}$ of the operator L which is being considered. As shown earlier, this condition has the form

$$\lfloor p(t)\varphi'(t)\rfloor_{t=0} = \theta\varphi(0).$$

We shall take the real parameter θ to be finite, so that $u(0; \lambda) \neq 0$, and we can therefore take

$$u(0; \lambda) = 1.$$

Further, we let $v(t; \lambda)$ denote a second solution of the homogeneous equation, normalized by the equalities

$$v(0; \lambda) = 0, \qquad [p(t)v'(t; \lambda)]_{t=0} = -1.$$

Here the kernel $K(t, s; \lambda)$ of the integral operator which is the resolvent of the operator $\tilde{L}$, *i.e.*, the Green's function, will have the form

$$K(t, s; z) = \begin{cases} u(s; z)[v(t; z) + m(z)u(t; z)], & (s \leq t) \\ u(t; z)[v(s; z) + m(z)u(s; z)], & (s > t) \end{cases} \qquad (\mathcal{I}(z) > 0), \tag{18}$$

where the function $m(z)$ is uniquely determined by the fact that

$$v(t; z)+m(z)u(t; z)\in L^2(0, \infty).$$

We now make the function $\varphi(t)$ traverse the set of all functions in $D(\tilde{L})$ which have compact support, and put

$$(\tilde{L}-zI)\varphi = f,$$

z (with $\mathscr{I}(z)>0$) being fixed. Then the set $\mathscr{N}$ traversed by the function $f(t)$ is dense in $L^2(0, \infty)$, since the deficiency index of the minimal operator L is (1, 1), by hypothesis. If $g(t)$ is an arbitrary function in $\mathscr{D}$, then

$$\int_0^\infty \varphi(t)\overline{g(t)}\,dt = \int_{-\infty}^\infty \Phi(\varphi; \lambda)\overline{\Phi(g; \lambda)}\,d\sigma(\lambda).$$

But since

$$\varphi(t) = \int_0^\infty K(t, s; z)f(s)\,ds,$$

and, by property 3° of direction functionals,

$$\Phi(\varphi; \lambda) = \frac{\Phi(f; \lambda)}{\lambda - z},$$

we have

$$\int_0^\infty \int_0^\infty K(t, s; z)f(t)\overline{g(t)}\,ds\,dt = \int_{-\infty}^\infty \frac{\Phi(f; \lambda)\overline{\Phi(g; \lambda)}}{\lambda - z}\,d\sigma(\lambda).$$

This equality holds for arbitrary $f, g\in\mathscr{D}$, since $\mathscr{N}$ is dense in $L^2(0, \infty)$. If we put

$$f(t) = g(t) = \begin{cases} 1/\delta & \text{for } 0\leq t\leq \delta, \\ 0 & \text{for } \quad t>\delta, \end{cases}$$

then the last equality can be written in the form

$$\frac{1}{\delta^2}\int_0^\delta \int_0^\delta K(t, s; z)\,dt\,ds = \int_{-\infty}^\infty \omega_\delta(\lambda)\frac{d\sigma(\lambda)}{\lambda - z}, \tag{19}$$

where

$$\omega_\delta(\lambda) = \left[\frac{1}{\delta}\int_0^\delta u(t; \lambda)\,dt\right]^2.$$

When $\delta\to 0$ the left-hand side of (19) has the limit

$$K(0, 0; z) = m(z).$$

Therefore

$$\lim_{\delta\to 0}\int_{-\infty}^{\infty}\omega_\delta(\lambda)\frac{d\sigma(\lambda)}{\lambda - z} = m(z),$$

which implies

$$\lim_{\delta\to 0}\int_{-\infty}^{\infty}\omega_\delta(\lambda)\frac{d\sigma(\lambda)}{\lambda^2+1} = \mathscr{I}(m(i)).$$

From this it is easy to deduce that the integral

$$\int_{-\infty}^{\infty}\frac{d\sigma(\lambda)}{\lambda^2+1}$$

exists and is equal to $\mathscr{I}(m(i))$. Therefore

$$\int_{-\infty}^{\infty}\left\{\frac{1}{\lambda - z}-\frac{1}{\lambda - i}\right\}d\sigma(\lambda) = m(z) - m(i),$$

and hence

$$m(z) = \mathscr{R}(m(i)) + \int_{-\infty}^{\infty}\frac{1+\lambda z}{\lambda - z}\frac{d\sigma(\lambda)}{\lambda^2+1}. \tag{20}$$

By applying the Stieltjes–Perron inversion formula (see §69) we now obtain

$$\frac{\sigma(\lambda - 0)+\sigma(\lambda+0)}{2} = \text{const.} + \lim_{y\to 0}\frac{1}{\pi}\int_0^{\lambda}\mathscr{I}(m(x+iy))\,dx.$$

It can be shown that this formula also holds when the boundary condition takes the form $\varphi(0) = 0$.

We shall now make some remarks about the case when the operator L has the deficiency index (2, 2). We shall discuss a class of self-adjoint extensions of this operator which were first investigated by H. Weyl [126].

Let L_θ denote that extension of the operator L (L may have the deficiency index (1, 1) or (2, 2)) which is determined by the boundary condition

$$[p(t)\varphi'(t)]_{t=0} = \theta\varphi(0), \tag{21}$$

so that $D(L_\theta)$ consists of all those functions $\varphi(t)\in D(L^*)$ which satisfy the condition (21). L_θ is clearly a symmetric extension of the operator L. If the deficiency index of L is (1, 1), then the operator L_θ turns out to be self-adjoint, and, by letting θ vary between $-\infty$ and $+\infty$, we obtain all the self-adjoint extensions of the operator L. But if the deficiency index of L is (2, 2), then the deficiency index of the operator L_θ will be (1, 1). Any self-adjoint extension $\tilde{L}_\theta$ of L_θ is an extension of the operator L. By letting θ vary between $-\infty$ and $+\infty$ and taking all the possible extensions $\tilde{L}_\theta$ for each value of θ, we obtain a certain class of self-adjoint extensions of the operator L. This class is characterized by the fact that one of the two boundary conditions (2) in §127 which determine the self-adjoint extensions takes the form (21); and the second of these conditions will evidently relate here only to the singular end-point $t=\infty$. Thus the class of self-adjoint extensions which has here been obtained is characterized by *separated boundary-conditions.*

The assumption made at the beginning of this section that the operator L has the deficiency index (1, 1) was used once only; namely, in proving property 2° of the direction functional. However, it is easy to see that this property 2° still holds even when the deficiency index is (2, 2), provided that we take as self-adjoint extensions $\tilde{L}$ only those extensions which are characterized by separated boundary-conditions. We shall denote extensions of this sort by $\tilde{L}_\theta$.

Thus all the above propositions remain true for the extensions $\tilde{L}_\theta$ of an operator L with the deficiency index (2, 2). In particular, it follows from Theorem 2 that the self-adjoint extensions $\tilde{L}_\theta$ have simple spectra.

It is easy to see that, just as in the case where the deficiency index is (1, 1), the kernel of the resolvent $\tilde{R}(z)$ of the operator $\tilde{L}_\theta$ (with $\theta \neq \infty$) is given by the formulae (18), in which the function $m(z)$ is determined uniquely by the choice of the particular self-adjoint extension $\tilde{L}_\theta$ of the operator L_θ. In particular, $K(0, 0; z) = m(z)$. Arguing further, as for the case of the deficiency index (1, 1), we arrive at formula (20). Since, by Theorem 2 of §128, the operator $\tilde{L}_\theta$ has a purely point spectrum with a unique limit-point at infinity, so the function $\sigma(\lambda)$ is a step-function, and formula (20) shows that $m(z)$ is a meromorphic function of z.

Correspondingly with the results of §113, the scalar product $(\tilde{R}(z)f, f)$, when $f(t) \in L^2(0, \infty)$ and z (with $\mathcal{I}(z) > 0$) are fixed, traverses a certain circle $C(f; z)$ when $\tilde{R}(z)$ runs over the set of all orthogonal resolvents of the operator L_θ. But it follows from formula (18) that the scalar product $(\tilde{R}(z)f, f)$ takes the form $\alpha + \beta m(z)$, where the constants α and β do not depend on the choice of the resolvent. Therefore, when the point $(\tilde{R}(z)f, f)$ runs over the circle $C(f; z)$, the point $w = m(z)$ also runs over a certain circle, which is called Weyl's *limit-circle.* When the deficiency index is (1, 1), and only in that case, Weyl's limit-circle degenerates into a point. The meaning of the terms *limit-circle* and *limit-point* introduced at the end of §127 now becomes clear. We shall not go into the method by which Weyl arrived at the idea of the limit-circle.

We observe that, although the circle $C(f; z)$ depends on the choice of both f and z, the corresponding limit-circle C depends only on z, but does not depend on f, because the function $m(z)$ does not depend on f.

Let us find an equation for the limit-circle. To do this, following M. G. Krein, we use the Hilbert functional relation for the resolvent of the operator $\tilde{L}_\theta$; from this we easily find that

$$K(t, s; z) - K(t, s; \bar{z}) = (z - \bar{z}) \int_0^\infty K(t, \xi; z) K(\xi, s; \bar{z}) \, d\xi.$$

Putting $t = s = 0$ in this equation and using (16), we obtain

$$m(z) - \overline{m(z)} = (z - \bar{z}) \int_0^\infty |v(\xi; z) + m(z) u(\xi; z)|^2 \, d\xi,$$

and so the equation of the limit-circle in the complex w-plane has the form

$$\int_0^\infty |v(\xi; z) + w u(\xi; z)|^2 \, d\xi = \frac{w - \bar{w}}{z - \bar{z}}.$$

In the case $\theta = \infty$ some modification of the argument is needed, but a similar result is obtained. We shall not go into the details.

In conclusion we touch briefly on the so-called inverse problem of spectral analysis.

As shown in Theorem 2, corresponding to every self-adjoint operator generated by the differential expression

$$-y'' + q(x) y$$

in $L^2(0, \infty)$, there is a spectral function $\sigma(\lambda)$, which generates the inversion formulae and the Parseval relation. It is a quite remarkable fact that the converse also holds: the differential operator of the form $-y'' + q(x)y$ with the associated boundary-conditions can be reconstructed unambiguously from the corresponding spectral function.

The problem of reconstructing a differential operator from its spectral function, and related topics, have acquired the name of the *inverse problem of spectral analysis.*

The inverse problem of spectral analysis for differential operators of the second order† can be regarded as the continuous analogue of the construction of orthogonal polynomials. Let us explain this point.

Suppose a spectral function $\sigma(\lambda)$ $(-\infty < \lambda < \infty)$ is given, satisfying conditions (which we need not enumerate here) under which it is uniquely associated with the equation

$$-y'' + q(x) y = \lambda y \qquad (x > 0) \tag{22}$$

† A complete solution of this problem and various generalizations have been given in the works of I. M. Gelfand and B. M. Levitan, M. G. Krein, and V. A. Marchenko. An exposition of the solutions of the inverse problem due to these authors can be found in the survey article, M. G. Gasimov and B. M. Levitan [31]: *see also* Naimark [90], Levitan and Sargsyan [72], and Marchenko [83].

together with the boundary condition

$$y'(0)+hy(0)=0. \tag{23}$$

The solution $\varphi(x;\lambda)$ of this equation such that $\varphi(0;\lambda)=1$ is an integral function of λ for each $x\geqslant 0$. From the equation (22) and the boundary condition (23) it follows that

$$\varphi(x;\lambda)=\cos(x\sqrt{\lambda})-\frac{h\sin(x\sqrt{\lambda})}{\sqrt{\lambda}}+\int_0^x \frac{\sin((x-\xi)\sqrt{\lambda})}{\sqrt{\lambda}}\,q(\xi)\varphi(\xi;\lambda)\,d\xi.$$

Hence, by putting $\lambda=s^2$, $t=\mathscr{I}(s)$, it is easy to obtain by a well-known method the estimate

$$|\varphi(x;\lambda)-\cos(x\sqrt{\lambda})|\leqslant\frac{e^{x|t|}}{|s|}\cdot\frac{Q(x)+|h|}{1-Q(x)/|s|},$$

which holds when

$$|s|>Q(x), \qquad \text{where} \qquad Q(x)=\int_0^x |q(\xi)|\,d\xi.$$

This estimate shows that $\varphi(x;\lambda)$, like $\cos(x\sqrt{\lambda})$, is an even, integral function of s, of exponential type in x, and that the difference $\varphi(x;\lambda)-\cos(x\sqrt{\lambda})$ belongs to $L^2(-\infty,\infty)$ on the real axis of s. Therefore the well-known Paley–Wiener theorem [97] can be applied to this difference to show that there is a function $K(x,\xi)$ such that

$$\varphi(x;\lambda)=\cos(x\sqrt{\lambda})+\int_0^x K(x,\xi)\cos(\xi\sqrt{\lambda})\,d\xi \tag{24}$$

and

$$\int_0^x |K(x,\xi)|^2\,d\xi<\infty.$$

The triangular kernel $K(x,\xi)$ $(0<\xi<x)$, which enables the family of functions $\{\cos(x\sqrt{\lambda})\}$ to be transformed into the family $\{\varphi(x;\lambda)\}$ (where $x\geqslant 0$ is a parameter), is analogous to a triangular matrix, by means of which a sequence $\{\lambda^k\}_0^\infty$ is transformed into a family of polynomials orthogonal relative to a given weight. The orthogonality relationship in the continuous case is expressed by the Parseval relation. Because of this analogy it is natural to call $K(x,\xi)$ the *orthogonalizing kernel*. The right-hand side of the relation (24), which transforms the family $\{\cos(x\sqrt{\lambda})\}$ into the family $\{\varphi(x;\lambda)\}$ is a particular case of a *transformation operator*. The general theory of these operators was constructed by V. A. Marchenko [82].

The function $q(x)$ in equation (22) is often called the *potential*, and equation (22) itself is called the (1-*dimensional*) *Schrödinger equation.* This terminology comes from quantum mechanics. In connection with certain problems in physics Bargman investigated degenerate orthogonalizing kernels

$$K(x, \xi) = \sum_{i=1}^{n} \alpha_i(x)\beta_i(\xi).$$

The potentials discovered in this way by Bargman are named after him and are used in the theory of scattering (see R. G. Newton [96]).

§130. Generalization to differential operators of any order

In §86 we saw that every self-adjoint operator A with a spectrum of multiplicity $r<\infty$ generates isometric mappings of H on to the space L_S^2 of vector functions. Under these mappings the operator A goes over into the operator of multiplication by the independent variable. We recall that the matrix distribution function $S(\lambda)$ is determined by the resolution of the identity E_λ for the operator A if we select some generating basis $g_1, g_2, \ldots, g_p$ $(r \leq p < \infty)$ of this operator, and put

$$S(\lambda) = [(E_\lambda g_j, g_k)]_{j,k=1}^{p}.$$

Further, the isometric mapping in question is defined by the formulae

$$\vec{\Phi}(\lambda) = Vf, \tag{1}$$

$$f = V^{-1}\vec{\Phi}(\lambda) = \text{l.i.m.} \int_{-M}^{N} \sum_{k=1}^{p} \Phi_k(\lambda)\, dE_\lambda g_k, \tag{2}$$

where the element f and the vector function $\vec{\Phi}(\lambda) = \{\Phi_1(\lambda), \ldots, \Phi_p(\lambda)\}$ traverse the spaces H and L_S^2 respectively.

We shall call (1) and (2) the *inversion formulae* associated with the self-adjoint operator A.

We point out that this definition is wider than that given in §129 for operators with a simple spectrum, since it does not require that the basis which generates the distribution function $S(\lambda)$ be minimal. In particular, by the definition in this section, with operators which have a simple spectrum there are associated inversion formulae defining an isometric mapping between H and L_S^2 in which the order of the matrix $S(\lambda)$ is greater than 1.

We shall now outline the derivation of the inversion formulae associated with a differential operator of arbitrary order.

Let L be the differential operator, with minimal domain, which is generated by a differential expression l of order $2n$ on the interval $(0, \infty)$ with the one singular end-point, and let $\tilde{L}$ be any self-adjoint extension of L.

We shall not now make any assumption about the deficiency index of the operator L.

On the manifold $\mathfrak{D}$ of those functions in $L^2(0, \infty)$ which have compact support we introduce the $2n$ direction functionals

$$\Phi_j(f; \lambda) = \int_0^\infty f(t) u_j(t; \lambda)\, dt \qquad (j = 1, 2, \ldots, 2n), \tag{3}$$

where $\{u_j(t; \lambda)\}_{j=1}^{2n}$ is the fundamental system of solutions of the equation

$$l[u] - \lambda u = 0$$

which satisfy the initial conditions

$$u_j^{[k-1]}(0; \lambda) = \begin{cases} 1 & (j = k), \\ 0 & (j \neq k), \end{cases} \qquad (j, k = 1, 2, \ldots, 2n).$$

It is easily verified that each of the functionals $\Phi_j(f; \lambda)$ has the properties 1° and 3° described in §129, but as regards property 2°, it now is formulated as follows.

2°. If, for some function $f \in \mathfrak{D}$ and for some real λ, the $2n$ equations

$$\Phi_j(f; \lambda) = 0 \qquad (j = 1, 2, \ldots, 2n)$$

hold, then the equation

$$(\tilde{L} - \lambda I)\varphi = f$$

has a solution in the class of functions with compact support.

Lemma 1.† *For any finite interval Δ_0 of the λ-axis there is a system of functions*

$$\psi_1(t), \psi_2(t), \ldots, \psi_{2n}(t)$$

belonging to $\mathfrak{D}$ and such that the determinant

$$D(\lambda) = \det\,[\Phi_j(\psi_k; \lambda)]_{j,k=1}^{2n}$$

does not vanish in the interval Δ_0.

The proof of this lemma is easily constructed on similar lines to that of the corresponding lemma in §129, by using the analyticity of the determinant $D(\lambda)$.

† This and the next proposition are due to M. G. Krein; see [59, 61].

Theorem 1. *Let Δ_0 be a finite interval, and let the system of functions $\psi_1, \psi_2, \ldots, \psi_{2n} \in \mathfrak{D}$ be such that*

$$\det [\Phi_j(\psi_k; \lambda)]_{j,k=1}^{2n} \neq 0 \qquad (\lambda \in \Delta_0).$$

Further, let E_λ be the resolution of the identity for the operator $\tilde{L}$, and let $g_1, g_2, \ldots, g_{2n}$ be a system of functions defined by the equalities

$$g_i(t) = \int_{\Delta_0} \sum_{k=1}^{2n} \Omega_{ik}(\lambda)\, dE_\lambda \psi_k,$$

where the $\Omega_{ik}(\lambda)$ are the elements of the matrix inverse to $[\Phi_j(\psi_k; \lambda)]_{j,k=1}^{2n}$. Then, for any function $f \in \mathfrak{D}$ and for any interval $\Delta \subseteq \Delta_0$, the formula

$$E(\Delta)f = \int_{\Delta} \sum_{j=1}^{2n} \Phi_j(f; \lambda)\, dE_\lambda g_j \tag{4}$$

holds.

Proof. The proof is entirely analogous to that of Theorem 1 in §129. Just as we did there, we replace the equality (4) to be proved by

$$E(\Delta)f = \int_{\Delta} \sum_{j=1}^{2n} F_j(f; \lambda)\, dE_\lambda \psi_j,$$

where

$$F_j(f; \lambda) = \sum_{k=1}^{2n} \Omega_{kj}(\lambda) \Phi_k(f; \lambda),$$

and we introduce the element

$$w_\mu = \int_a^\mu dE_\lambda f - \int_a^\mu \sum_{j=1}^{2n} F_j(f; \lambda)\, dE_\lambda \psi_j,$$

which proves to be continuously dependent on μ. The inequality (8) of §129 will now have the form (for $\delta \geqslant 0$)

$$\|w_{\mu+\delta} - w_\mu\| \leqslant \left\| E(\Delta_{\mu+0,\delta}) \left[f - \sum_{j=1}^{2n} F_j(f; \mu)\psi_j \right] \right\|$$
$$+ \left\| E(\Delta_{\mu+0,\delta}) \sum_{j=1}^{2n} F_j(f; \mu)\psi_j - \int_{\mu+0}^{\mu+\delta} \sum_{j=1}^{2n} F_j(f; \lambda)\, dE_\lambda \psi_j \right\|.$$

The terms on the right in this inequality are estimated in the same way as in §129, and from these estimates the theorem follows.

Theorem 2. *If L, $\tilde{L}$, and $u_j(f; \lambda)$ have the meanings defined at the beginning of this section, then there is a matrix distribution function*

$S(\lambda)=[\sigma_{ik}(\lambda)]_{i,k=1}^{n}$, *called the spectral matrix, by means of which the formulae*

$$\Phi_j(\lambda)=\text{l.i.m.}\int_{-M}^{N} f(t)u_j(t;\lambda)\,dt \qquad (j=1,2,\ldots,2n),$$

$$f(t)=\text{l.i.m.}\int_{-M}^{N}\sum_{i,k=1}^{n}\Phi_i(\lambda)u_k(t;\lambda)\,d\sigma_{ik}(\lambda)$$

establish an isometric mapping of the space $L^2(0,\infty)$ *on to the space* L_S^2 *of vector functions.*

Proof. We put $\Delta=\Delta_0$ in the formula (4) and replace the system of elements $\{g_j\}_{j=1}^{2n}$ by the system $\{g_j^{(0)}\}=\{E(\Delta_0)g_j\}$ $(j=1,2,\ldots,2n)$, and then divide the λ-axis into finite elements Δ_k $(\pm k=0,1,2,\ldots)$, and for each of these elements we select a system of elements $\{g_j^{(k)}\}_{j=1}^{2n}$, just as the system $\{g_j^{(0)}\}_{j=1}^{2n}$ was selected for the interval Δ_0. By putting

$$g=\sum_{k=-\infty}^{\infty} g_i^{(k)} \qquad (i=1,2,\ldots,2n)$$

we obtain, as in §129, a system of (improper) elements $g_1, g_2, \ldots, g_{2n}$ which, for any finite interval Δ, satisfy the condition

$$E(\Delta)g_i=\sum_k E(\Delta)g_i^{(k)}$$

(where the series on the right contains only a finite number of terms).

Further, for any finite interval Δ of the λ-axis we obtain the formula

$$E(\Delta)f=\int_{\Delta}\sum_{j=1}^{2n}\Phi_j(f;\lambda)\,dE_\lambda g_j,$$

and, for $\Delta=(-\infty,\infty)$,

$$f=\text{l.i.m.}\int_{-M}^{N}\sum_{j=1}^{2n}\Phi_j(f;\lambda)\,dE_\lambda g_j.$$

Since the manifold $\mathcal{D}$ is dense in $L^2(0,\infty)$, it follows from the last formula that the multiplicity of the spectrum of the operator $\tilde{L}$ does not exceed $2n$, and that $g_1, g_2, \ldots, g_{2n}$ is a generating basis.

For $f\in\mathcal{D}$, the Parseval relation takes the form

$$\int_0^\infty |f(t)|^2\,dt=\int_{-\infty}^{\infty}\sum_{j=1}^{2n}\Phi_j(f;\lambda)\overline{\Phi_k(f;\lambda)}\,d\sigma_{jk}(\lambda).$$

The rest of the argument in §129 carries over without difficulty to the present case, and so Theorem 2 is proved.

We point out that, in the case of separated boundary conditions (and, in particular, always when the deficiency index of the minimal operator is (n, n), since in that case all the boundary conditions are concentrated at the origin), the inversion formulae can be constructed by means of a spectral matrix of order n (and not $2n$). To do this, we have to take, instead of the $2n$ functions $u_j(t;\lambda)$, n linear combinations of them, $\tilde{u}_k(t;\lambda)$, which are linearly independent and which satisfy n boundary-conditions at the origin, and then consider a system of n direction-functionals

$$\tilde{\Phi}_k(f;\lambda)=\int_0^\infty f(t)\tilde{u}_k(t;\lambda)\,dt \qquad (k=1,2,\ldots,n).$$

§131. Analysis of the nature of the spectrum of a differential operator by the decomposition method

In those cases where the inversion formulae established in §129 and §130 can actually be constructed, the spectrum of the differential operator $\tilde{L}$ is found as the set of points of growth of the spectral function $\sigma(\lambda)$ of §129 (or of the matrix spectral function $S(\lambda)$ of §130). However, such cases are rarely encountered (examples will be given in §132). It is therefore of concern to analyze the various properties of the spectrum $\mathcal{S}(\tilde{L})$ by using the given information about the behaviour of the coefficients in the differential expression l, without requiring to know the spectral function and the inversion formulae. The topics arising from this problem constitute what is known as *qualitative spectral analysis of singular differential operators.*

In this section we shall consider some examples of the problems of spectral analysis by using the decomposition method.† We shall restrict our attention, for the sake of brevity, to the simplest (Schrödinger) differential expression

$$l[y]=-y''+q(x)y \tag{1}$$

on the half-line $x>0$ with the one singular end-point $x=\infty$.

The spectra of different self-adjoint extensions $\tilde{L}$ of the operator L with minimal domain generated by the differential expression (1) are distinct, but those of their properties which we shall be considering here do not depend on the particular extension chosen. Among properties of

† A brief account of this method is given in the paper by I. M. Glazman [36]. The monograph [37] by Glazman gives a systematic analysis of the nature of the spectrum of 1-dimensional and n-dimensional operators by the decomposition method. The theorems given here without proof are treated in detail in that monograph.

this sort are:

1° whether a given point λ belongs to $\mathscr{C}(\tilde{L})$;

2° whether there is an infinite set of points of the spectrum in an arbitrarily small one-sided neighbourhood of a point λ, and

3° whether there is a condensation point of the spectrum $\mathscr{S}(\tilde{L})$ at the point $\lambda = -\infty$.

The fact that these properties are independent of the particular extension chosen follows from Theorem 1 of §105, Theorem 1 of §82, and Theorem 2 of §107. We shall call them the *singular properties* of the spectrum $\mathscr{S}(\tilde{L})$, since the spectrum of a regular differential operator certainly does not have them.

Turning to the description of the decomposition method, we let L_0 denote the operator defined by the equality

$$L_0\varphi = \tilde{L}\varphi$$

on all those functions $\varphi \in D(\tilde{L})$ such that at a given point $\gamma > 0$ (called the *point of decomposition*), the conditions

$$\varphi(\gamma) = \varphi'(\gamma) = 0$$

are satisfied. Clearly, the operator L_0 splits up into the orthogonal sum

$$L_0 = L_r \oplus L_\gamma$$

of the operators L_r and L_γ which are generated by the same differential expression l in the spaces $L^2(0, \gamma)$ and $L^2(\gamma, \infty)$ respectively. By extending these two operators into self-adjoint operators $\tilde{L}_r$ and $\tilde{L}_\gamma$, we construct the operator

$$M = \tilde{L}_r \oplus \tilde{L}_\gamma .$$

The given operator $\tilde{L}$ and the constructed operator M now turn out to be different self-adjoint extensions of the same 'decomposed' operator L_0. Since the operator L_0 obviously has finite deficiency numbers in $L^2(0, \infty)$, and since the operator $\tilde{L}_r$ is regular, it follows, by Theorem 1 in §105, that

$$\mathscr{C}(\tilde{L}) = \mathscr{C}(M) = \mathscr{C}(\tilde{L}_r) \cup \mathscr{C}(\tilde{L}_\gamma) = \mathscr{C}(\tilde{L}_\gamma).$$

It is natural to expect that the other singular properties of the spectra of the operators $\tilde{L}$ and $\tilde{L}_\gamma$ will be the same. That this is so can be proved by means of the following

Lemma. *Let $n(\alpha, \beta; A)$ denote the number of points of the spectrum of an operator A which lie in the interval (α, β). Then, for any α and β $(-\infty \leq \alpha < \beta < \infty)$ the numbers $n(\alpha, \beta; \tilde{L})$ and $n(\alpha, \beta; \tilde{L}_\gamma)$ are either both finite or both infinite.*

Proof. If $n(\alpha, \beta; \tilde{L})<\infty$ (resp.$=\infty$), then by Theorem 1 of §82 we shall have $n(\alpha, \beta; M)<\infty$ (resp.$=\infty$). (If $\alpha=-\infty$, then we have to use Theorem 2 of §82 instead of Theorem 1.) Further, we have

$$\mathscr{S}(M)=\mathscr{S}(\tilde{L}_r)\cup\mathscr{S}(\tilde{L}_\gamma),$$

where the number of points of the spectrum $\mathscr{S}(\tilde{L}_r)$ which lie in the interval (α, β) is always finite, since the spectrum of a regular operator is discrete and bounded below.

It suffices to verify the latter statement for the case of the boundary conditions $y(0)=y(\gamma)=0$. Let

$$Q(x)=\int_0^x q(t)\,dt, \qquad \text{and let} \qquad |Q(x)|\leq\mathcal{M}.$$

Then

$$(\tilde{L}_r y, y)=\int_0^\gamma(|y'|^2+q(x)\,|y|^2)\,dx=\|y'\|^2-\int_0^\gamma Q(x)\frac{d}{dx}|y|^2\,dx$$
$$\geq\|y'\|^2-2\mathcal{M}\,\|y'\|\,.\,\|y\|\geq-\mathcal{M}^2\,\|y\|^2,$$

and this means that the operator $\tilde{L}_r$ is semi-bounded below.

Hence, if $n(\alpha, \beta; M)<\infty$ (resp.$=\infty$), then also $n(\alpha, \beta; \tilde{L}_\gamma)<\infty$ (resp.$=\infty$). Q.E.D.

Theorem 1. The singular properties of the spectra $\mathscr{S}(\tilde{L})$ and $\mathscr{S}(\tilde{L}_\gamma)$ are the same.

Proof. The property $\mathscr{C}(\tilde{L})=\mathscr{C}(\tilde{L}_\gamma)$ has already been proved. The singular properties 2° and 3° follow from the lemma if we put respectively

$$\alpha=\lambda-\varepsilon,\quad \beta=\lambda \quad (\varepsilon\searrow 0), \quad\text{and}\quad \alpha=-\infty,\quad \beta<\infty \quad (|\beta|\to\infty).$$

Q.E.D.

The decomposition method is based on this Theorem 1, and also on Theorems 1 and 2 of §82, Theorem 4 of §93, and Lemma 2 of §94. In applying the theorems from §82 the linear manifold G is formed as the linear envelope of a sequence of functions of compact support whose supports are disjoint.

If $\lambda\in\mathscr{C}(\tilde{L})$, then either the point λ is a two-sided limit-point of the set $\mathscr{S}(\tilde{L})$, or it is a one-sided left limit-point (resp. right limit-point) of this set. Depending on which one of these cases occurs, let us agree to classify any point $\lambda\in\mathscr{C}(\tilde{L})$ under one of these three possible *types*.

From Theorem 1 we have immediately the following proposition—which can be generalized without difficulty for any differential expression of the form (1) in §123.

Theorem 2.

1. *The continuous part $\mathscr{C}(\tilde{L})$ of the spectrum,*
2. *the type of each point in $\mathscr{C}(\tilde{L})$, and*
3. *the semi-boundedness below of $\mathscr{S}(L)$ or the non-semi-boundedness below of $\mathscr{S}(L)$*

—— *all these three things do not depend on the behaviour of the potential $q(x)$ at a finite distance.*

Thus all the singular properties of the spectrum of the operator $\tilde{L}$ are determined by the behaviour of the potential $q(x)$ at infinity. In terms of perturbations, Theorem 2 means that any perturbation $\eta(x)$, of the potential $q(x)$, which has compact support does not change the singular properties of the spectrum. (It is assumed here, and below, that the perturbation $\eta(x)$ is, like $q(x)$, absolutely integrable on any finite interval.) If the perturbation $\eta(x)$ does not have compact support but $\eta(x)\to 0$ as $x\to\infty$, then the type of the points in $\mathscr{C}(\tilde{L})$ may change (see Theorem 5 below), but property 3° is preserved. The following theorem shows that property 1° is also preserved.

Theorem 3. *If $\tilde{L}'$ is a self-adjoint differential operator generated in* $L^2(0,\infty)$ *by the differential expression*

$$-y''+q(x)y+\eta(x)y$$

and if

$$\lim_{x\to\infty}\eta(x)=0, \tag{2}$$

then

$$\mathscr{C}(\tilde{L}')=\mathscr{C}(\tilde{L}). \tag{3}$$

Proof. By splitting the operators $\tilde{L}'$ and $\tilde{L}$ at some point γ we obtain

$$\mathscr{C}(\tilde{L})=\mathscr{C}(\tilde{L}_\gamma) \qquad \text{and} \qquad \mathscr{C}(\tilde{L}')=\mathscr{C}(\tilde{L}'_\gamma). \tag{4}$$

Then choosing γ so that $|\eta(x)|<\delta$ when $x>\gamma$, and using Lemma 2 of §94, we deduce that in a δ-neighbourhood of each point $\lambda\in\mathscr{C}(\tilde{L}_\gamma)$ there are points of $\mathscr{C}(\tilde{L}'_\gamma)$, and conversely, in a δ-neighbourhood of each point $\lambda\in\mathscr{C}(\tilde{L}'_\gamma)$ there are points of $\mathscr{C}(\tilde{L}_\gamma)$. Since the number δ can be arbitrarily small, the relation (3) follows from (4). Q.E.D.

In the case where $q(x)\geqslant q_0>0$, Theorem 3 generalizes to perturbations which are *relatively* small, *i.e.*, instead of (2) we require that

$$\lim_{x\to\infty}\frac{\eta(x)}{q(x)}=0.$$

This result is due to M. Sh. Birman.

As a particular case, Theorem 3 with $q(x)=0$ and the results of §89 imply

Theorem 4. *If* $\lim_{x\to\infty} q(x)=0$, *then* $\mathscr{C}(\tilde{L})=[0,\infty)$.

Among more general criteria which ensure that $\mathscr{C}(\tilde{L})=[0,\infty)$ we may mention the condition

$$\lim_{x\to\infty}\int_x^{x+\omega}|q(t)|\,dt=0$$

for some $\omega>0$ (and therefore also for any $\omega>0$). This condition holds, in particular, if the potential $q(x)\in L^2(0,\infty)$.

We now investigate the type of the point $\lambda=0$ under the conditions of Theorem 4. If $q(x)\geqslant 0$ for large x, then, from the lemma of this section, it follows that the negative part of the spectrum $\mathscr{S}(\tilde{L})$ consists only of a finite number of eigenvalues, and so the point $\lambda=0$ is not a right-limit-point of the spectrum $\mathscr{S}(\tilde{L})$. On the other hand, if, for large x, the potential were constant, say $q(x)\equiv-\varepsilon$, then we would have $\mathscr{C}(\tilde{L})=[-\varepsilon,\infty)$, and therefore, however small $\varepsilon>0$ may be, the point $\lambda=0$ would be a right-limit-point for $\mathscr{S}(\tilde{L})$. It is therefore natural to expect that, under the conditions of Theorem 4, the type of the point $\lambda=0$ will depend on 'the degree of negativity' of the potential $q(x)$ as $x\to\infty$. The 'boundary' between the potentials which determine the two different types of the point $\lambda=0$ is fixed by the following theorem. The existence of the limit

$$\lim_{x\to\infty} q(x)$$

is not assumed in the hypothesis of this theorem.

Theorem 5. *If*

$$q(x)\geqslant-\frac{1}{4x^2} \tag{5}$$

for large x, then the negative part of $\mathscr{S}(\tilde{L})$ *consists of only a finite number of eigenvalues. But if, for some* $\delta>0$,

$$q(x)<-\frac{1+\delta}{4x^2} \tag{6}$$

when x is large, then the negative semi-axis contains an infinite number of points of the spectrum $\mathscr{S}(\tilde{L})$.

Proof. If (5) holds for $x \geqslant \gamma$, then, for any function $y \in D(\tilde{L})$ which has compact support and whose support lies to the right of the point γ, we have

$$(\tilde{L}y, y) = \int_\gamma^\infty \{|y'|^2 + q(x)\,|y|^2\}\,dx$$

$$\geqslant \int_\gamma^\infty \left\{|y'|^2 - \frac{1}{4x^2}|y|^2\right\} dx = \int_\gamma^\infty \left|y' - \frac{1}{2x}y\right|^2 dx \geqslant 0.$$

By splitting the operator $\tilde{L}$ at the point γ and using Theorem 1 of §107 we easily obtain the proof of the first part of the theorem.

To prove the second part of the theorem under the condition (6), we construct an infinite sequence of functions $y \in D(\tilde{L})$ of compact support, which have disjoint supports and are such that $(\tilde{L}y, y) < 0$. Since this same inequality will clearly also be satisfied by any element of the infinite-dimensional sub-space G spanned by the constructed sequence, it will therefore follow, by Theorem 2 of §82, that the negative semi-axis will contain an infinite number of points of the spectrum $\mathscr{S}(\tilde{L})$, as was to be shown.

So, let the condition (6) be satisfied for $x \geqslant \gamma$, and let $y \in D(\tilde{L})$ be any function of compact support, whose support is to the right of the point γ. Putting

$$x = e^s, \quad \frac{y(x)}{\sqrt{x}} = z(s), \quad \text{and} \quad \alpha = \log \gamma,$$

we have

$$(\tilde{L}y, y) \leqslant \int_\gamma^\infty \left\{\left|y' - \frac{y}{2x}\right|^2 - \frac{\delta}{4x^2}|y|^2\right\} dx$$

$$= \int_\alpha^\infty \left\{|z'|^2 - \frac{\delta}{4}|z|^2\right\} ds. \tag{7}$$

Hence it is clear that, if we put

$$z(s) = \varphi_N(s - a),$$

where

$$\varphi_N(s) = \begin{cases} \sin^2 s & \text{for} \quad 0 \leqslant s \leqslant \pi/2, \\ 1 & \text{for} \quad \pi/2 \leqslant s \leqslant \pi/2 + N, \\ \sin^2(s - N) & \text{for} \ \pi/2 + N \leqslant s \leqslant \pi + N, \\ 0 & \text{for} \quad s < 0 \text{ or } s > \pi + N \end{cases}$$

and $a > \alpha$ is arbitrary, then for the corresponding function $y(x)$ we shall have $(\tilde{L}y, y) < 0$ when N (which is independent of a) is sufficiently large.

Having chosen such an N and then giving a in succession the values

$$a_k = \alpha + (N + 2\pi)k, \qquad k = 1, 2, \ldots,$$

we obtain a sequence of functions

$$z_k(s) = \varphi_N(s - a_k),$$

to which will correspond the required sequence $y_k(x)$. Hence the theorem.

The inequalities (5) and (6) are the same as the classical criteria, due to A. Kneser, for non-oscillation and oscillation of solutions of the differential equation

$$y'' - q(x)y = 0. \tag{8}$$

We shall say that the differential expression (1) is *oscillatory* if the solution of the equation (8) has an infinite number of zeros on the semi-axis $x > 0$, and *non-oscillatory* otherwise. We recall that the zeros of any two linearly independent solutions of equation (8) interlace, and so, if one solution has an infinite number of zeros, then any other solution will also have an infinite number of zeros.

The following general proposition establishes a connection between the negative part of the spectrum $\mathscr{S}(\tilde{L})$ and the oscillation of the differential expression (1).

Theorem 6. The differential expression (1) *is oscillatory if and only if the negative part of the spectrum of the operator* $\tilde{L}$ *is an infinite set.*

Proof. Let a solution $y(x)$ of equation (8) have an infinite number of zeros $\alpha_1 < \beta_1 < \alpha_2 < \beta_2 < \ldots$, so that

$$y(\alpha_k) = y(\beta_k) = 0 \qquad (k = 1, 2, 3, \ldots). \tag{9}$$

In the interval (α_k, β_k) the function $y(x)$ can be regarded as an eigenfunction of the problem $l[y] = \lambda y$ with the boundary conditions (9), this eigenfunction corresponding to the eigenvalue $\lambda = 0$. From variational principles it follows that the eigenvalue problem $l[y] = \lambda y$ with the boundary conditions

$$y(\alpha_k) = y(\beta_k') = 0, \qquad \text{where} \quad \beta_k' > \beta_k,$$

must have a negative eigenvalue λ_k. Let $\varphi_k(x)$ be the corresponding eigenfunction, so that

$$\int_{\alpha_k}^{\beta_k'} l[\varphi_k]\bar{\varphi}_k \, dx = \lambda_k \int_{\alpha_k}^{\beta_k'} |\varphi_k|^2 \, dx < 0,$$

i.e.,

$$\int_{\alpha_k}^{\beta_k'} \{|\varphi_k'|^2 + q(x)\,|\varphi_k|^2\}\,dx < 0. \tag{10}$$

We now continue $\varphi_k(x)$ on to the whole semi-axis $x > 0$, smoothing it in the neighbourhood of the end-points α_k and β_k' down to zero, so that the resulting function $\tilde{\varphi}(x)$ belongs to $D(\tilde{L})$. This smoothing is easily carried out in such a way that the interval $[\alpha_k, \beta_k']$ is kept as the support of the function $\tilde{\varphi}_k(x)$, which is a function of compact support, and so that the inequality (10) still holds; the latter can now be presented in the form

$$(\tilde{L}\tilde{\varphi}_k, \tilde{\varphi}_k) < 0 \qquad (k = 1, 2, 3, \ldots).$$

Since the supports of the constructed functions are disjoint, it follows that, for any function $\varphi(x)$ in the infinite-dimensional envelope G of the functions $\tilde{\varphi}_k$ $(k = 1, 2, 3, \ldots)$, the inequality $(\tilde{L}\varphi, \varphi) < 0$ will also hold, and hence, by Theorem 2 of §82, it follows that the set of points in the negative part of the spectrum of the operator $\tilde{L}$ is infinite.

Conversely, suppose that the set $\mathcal{S}(\tilde{L}) \cap (-\infty, 0)$ is infinite, let $y(x)$ be a solution of equation (8), and let N be an arbitrarily prescribed positive number. It is required to prove that $y(x)$ has a root greater than N.

Since the set $\mathcal{S}(\tilde{L}) \cap (-\infty, 0)$ is infinite, we conclude from the lemma that, for any γ, the set $\mathcal{S}(\tilde{L}_\gamma) \cap (-\infty, 0)$ is infinite.

By Theorem 2 of §82 there is an infinite-dimensional manifold of functions $\varphi(x)$ in $D(\tilde{L}_\gamma)$ such that $(\tilde{L}_\gamma\varphi, \varphi) < 0$. Since $\tilde{L}_\gamma$ is a finite-dimensional extension of the differential operator L_γ with minimal domain, the manifold of functions $\varphi \in D(L_\gamma)$ such that $(L_\gamma\varphi, \varphi) < 0$ is also infinite-dimensional (although for us the only important thing is that this manifold is not empty for any $\gamma > 0$).

Let $\varphi(x)$ be one of these functions. Since the operator L_γ is the closure of the operator $\mathring{L}_\gamma$ which coincides with L_γ on those functions in $D(L_\gamma)$ which have compact support, there is a function $\mathring{\varphi}(x) \in D(L_\gamma)$ with compact support and such that

$$(L_\gamma\mathring{\varphi}, \mathring{\varphi}) < 0. \tag{11}$$

Suppose that the function $\mathring{\varphi}(x)$ of compact support is equal to 0 outside an interval (a, b). Then the inequality (11) implies that the regular differential operator generated by the expression (1) with the boundary conditions $y(a) = y(b) = 0$ has at least one negative eigenvalue λ. When b decreases continuously, this eigenvalue $\lambda = \lambda(b)$ varies continuously, and when $b \to a$ it increases without limit. Therefore there is a value b^* $(a < b^* < b)$ such that $\lambda(b^*) = 0$. The corresponding eigenfunction $\varphi^*(x)$ satisfies the inequality (11) and the boundary conditions

$$\varphi^*(a) = \varphi^*(b^*) = 0.$$

But then any solution $y(x)$ must have a root in the interval $[a, b]$. Since we can take $\gamma > N$ and $a \geq \gamma$, the theorem is proved.

Of other criteria for non-oscillation and oscillation, we mention the following theorem, in the hypothesis of which it is assumed that $q(x)$ is non-positive for large x.

If

$$\limsup_{x\to\infty} x\int_x^\infty |q(t)|\,dt < \tfrac{1}{4},$$

the differential expression (1) *is non-oscillatory; and if*

$$\limsup_{x\to\infty} x\int_x^\infty |q(t)|\,dt > 1$$

it is oscillatory. (See Hille [45].)

We now discuss criteria for the semi-boundedness and discreteness of the whole spectrum $\mathcal{S}(\tilde{L})$.

Theorem 7. If

$$\lim_{x\to\infty} q(x) = +\infty,$$

then the spectrum of the operator $\tilde{L}$ is bounded below and discrete.

Proof. That the operator $\tilde{L}$, and therefore also its spectrum, is bounded below is obvious.

To prove that $\mathcal{S}(\tilde{L})$ is discrete, we first specify arbitrarily an $N > 0$, and then choose a point of decomposition γ so that $q(x) > N$ for $x \geq \gamma$. Then, for any function $y(x) \in D(\tilde{L}_\gamma)$ with compact support, we have

$$(\tilde{L}_\gamma y, y) = \int_\gamma^\infty (|y'|^2 + q(x)\,|y|^2)\,dx > N\int_\gamma^\infty |y|^2\,dx,$$

and so the intersection $\mathcal{C}(\tilde{L}_\gamma) \cap (-\infty, N)$ is empty, and therefore the intersection $\mathcal{C}(\tilde{L}) \cap (-\infty, N)$ is empty. Hence, since N is arbitrary, it follows that $\mathcal{C}(\tilde{L})$ is empty. Q.E.D.

A more general criterion, due to A. M. Molchanov, is as follows.

If the potential $q(x)$ is bounded below, then the spectrum of the operator L is discrete if and only if the mean value of the potential over any finite *interval tends to infinity, i.e., if*

$$\lim_{x\to\infty}\int_x^{x+\omega} q(t)\,dt=+\infty \qquad \textit{for any } \omega>0.$$

According to Theorem 3, in cases where there is a finite limit $\lim_{x\to\infty} q(x)=q_\infty$, *i.e.*, where the oscillation of the potential at infinity

$$\omega=\limsup_{x\to\infty} q(x)-\liminf_{x\to\infty} q(x)$$

is zero, the continuous spectrum of the operator $\tilde{L}$ coincides with the half-line $[q_\infty, \infty)$. But now the question arises: if the oscillation ω of the potential at infinity is finite and non-zero, how will this affect the continuous spectrum? It is natural to expect that the effect will somehow be governed by the value of ω. In fact, the following theorem holds.

Theorem 8. *The length of the gaps in the continuous part of the spectrum of the operator $\tilde{L}$ does not exceed ω.*

Proof. We put

$$\sigma=\liminf_{x\to\infty} q(x), \qquad \rho=\limsup_{x\to\infty} q(x),$$

and introduce the differential expression

$$l[y]=-y''+Q(x)y \tag{12}$$

with the potential

$$Q(x)=q(x)-(\sigma+\rho)/2.$$

Then

$$\limsup_{x\to\infty} Q(x)=\mathcal{M}.$$

where

$$\mathcal{M}=\omega/2=(\rho-\sigma)/2,$$

and it suffices to show that the length of the gaps in the continuous part of the spectrum of the operator $\tilde{L}$ generated by the differential expression (12) does not exceed $2\mathcal{M}$.

It is clear that there are no points of the continuous part of the spectrum of the operator $\tilde{L}$ to the left of the point $-\mathcal{M}$. Now suppose that $\lambda_0>0$ and $\delta>\mathcal{M}$, but that the interval $[\lambda_0-\delta, \lambda_0+\delta]$ contains no points of $\mathscr{C}(\tilde{L})$. We shall show that this supposition leads to a contradiction.

Since $\mathscr{C}(\tilde{L}) = \mathscr{C}(\tilde{L}_\gamma)$, it follows from the assumption just made that, for any γ, the intersection $\mathscr{C}(\tilde{L}_\gamma) \cap (\lambda_0 - \delta, \lambda_0 + \delta)$ must be empty. If we now choose the point of decomposition γ so that $|Q(x)| \leq \delta_0 < \delta$ for $x > \gamma$, then, using Lemma 2 of §94, we conclude that the point λ_0 does not belong to the continuous part of the spectrum of the operator of second-order differentiation P^2 in $L^2(\gamma, \infty)$, and this is absurd, because the continuous part of that spectrum covers the half-line $\lambda \geq 0$ completely.

Q.E.D.

With some additional restrictions, Theorem 8 can be extended considerably. For example, if the function $q(x)$ is bounded and uniformly continuous on the semi-axis $x > 0$, then the length of those gaps in $\mathscr{C}(\tilde{L})$ which lie to the right of the point λ tends to zero, by the Hartman–Putnam theorem. This is also true for any locally integrable, periodic potential.

If the potential $q(x)$ has finite oscillation at infinity, then its perturbation by a term $\eta(x)$ which tends to 0 as $x \to \infty$ does not change $\mathscr{C}(\tilde{L})$, but it may carry the discrete spectrum into the gaps of the continuous spectrum $\mathscr{C}(\tilde{L})$. The number of points of the discrete spectrum thus carried into a gap may be finite or infinite. The question as to which of these two cases occurs reminds us of the problem of finding conditions for non-oscillation or oscillation. However, this question is actually very much more complicated, and so far it has not been investigated, except for the case of the Schrödinger operator with a periodic potential (this is also called the *Hill operator*). We state the relevant results.

Theorem 9. (see Rofe-Beketov [107])
If

$$\int_0^\infty x\,|\eta(x)|\,dx < \infty \qquad \textit{resp.} \qquad \int_{-\infty}^\infty |x\eta(x)|\,dx < \infty,$$

then the discrete spectrum carried by a perturbation $\eta(x)$ into each gap of the continuous spectrum $\mathscr{C}(\tilde{L})$ of the Hill operator considered on the semi-axis (resp. on the whole axis) is finite, and sufficiently remote gaps contain no more than one (resp. two) eigenvalues apiece.

Theorem 10. (see Rofe-Beketov [110])
If, in addition to the hypothesis of Theorem 9, the condition

$$\int_{-\infty}^\infty \eta(x)\,dx \neq 0$$

is also satisfied, then every sufficiently remote gap contains precisely one eigenvalue of the perturbed Hill operator on the whole axis.

§132. Examples

1. Let us pose the problem of finding a potential $q(x)$ such that the equation

$$-y''+q(x)y=\lambda y \tag{1}$$

shall have as one of its solutions the function

$$\varphi(x;\lambda)=\cos(x\sqrt{\lambda})+T(x)\frac{\sin(x\sqrt{\lambda})}{\sqrt{\lambda}},$$

where $T(x)$ does not depend on λ. Thus (*cf.* the end of §129) we are trying to determine the potential $q(x)$ from the condition that the orthogonalizing kernel shall be degenerate and have the form

$$K(x,\xi)=T(x).$$

By substituting $\varphi(x;\lambda)$ into the equation (1), we obtain the two relations

$$2T'(x)=q(x), \qquad T''(x)=q(x)T(x),$$

from which it follows that

$$T(x)=-A\tanh(Ax+B), \qquad q(x)=-\frac{2A^2}{\cosh^2(Ax+B)},$$

where A and B are arbitrary real constants. Since a change in the sign of A is equivalent to a change in the sign of B, let us assume that $A\geqslant 0$. At the point $x=0$ the function $\varphi(x;\lambda)$ satisfies the boundary condition

$$y'(0)=T(0)y(0), \qquad \varphi(0;\lambda)=1.$$

It is easy to verify that the second solution of the equation with the potential just found is the function

$$\theta(x;\lambda)=\sqrt{\lambda}\sin(x\sqrt{\lambda})-T(x)\cos(x\sqrt{\lambda});$$

this satisfies, at the point $x=0$, the condition

$$\theta(0;\lambda)=-T(0), \qquad \theta'(0;\lambda)=\lambda-T'(0).$$

Instead of the function $\theta(x;\lambda)$ we shall take, as the second solution of the equation under consideration, the function

$$\psi(x;\lambda)=\frac{1}{A^2+\lambda}\{T(0)\varphi(x;\lambda)+\theta(x;\lambda)\},$$

for which

$$\psi(0;\lambda)=0, \qquad \psi'(0;\lambda)=1.$$

We now deal with the application of the general theory to the differential expression

$$l=-\frac{d^2}{dx^2}+q(x) \qquad (0\leqslant x<\infty).$$

The deficiency index of the operator L is (1, 1).

Let the boundary condition characterizing a self-adjoint extension of the operator have the form

$$y'(0)=hy(0). \tag{2}$$

It will be convenient to put $h=T(0)+\gamma$. We shall have to consider separately the boundary condition

$$y(0)=0. \tag{2'}$$

Corresponding to the boundary condition (2) there is a fundamental system of solutions, *viz.*,

$$\begin{aligned} u(x;\lambda)&=\varphi(x;\lambda)+\gamma\psi(x;\lambda)\\ &=\frac{A^2+\lambda+\gamma[T(0)-T(x)]}{A^2+\lambda}\cos(x\sqrt{\lambda})+\\ &\quad+\frac{(A^2+\lambda)T(x)+\gamma T(0)T(x)+\gamma\lambda}{A^2+\lambda}\cdot\frac{\sin(x\sqrt{\lambda})}{\sqrt{\lambda}}, \end{aligned} \tag{2_1}$$

$$v(x;\lambda)=-\psi(x;\lambda)=\frac{T(x)-T(0)}{A^2+\lambda}\cos(x\sqrt{\lambda})-\frac{T(0)T(x)+\lambda}{A^2+\lambda}\cdot\frac{\sin(x\sqrt{\lambda})}{\sqrt{\lambda}}. \tag{2_2}$$

To determine the function $m(\zeta)$ $\quad(\mathscr{I}(\zeta)>0)$, we form the sum

$$\begin{aligned} &v(x;\zeta)+m(\zeta)u(x;\zeta)=\exp[ix\sqrt{\zeta}]\{\cdots\}\\ &\quad+\exp[-ix\sqrt{\zeta}]\left\{\frac{T(x)-T(0)}{2(A^2+\zeta)}-\frac{i}{2}\frac{\zeta+T(0)T(x)}{(A^2+\zeta)\sqrt{\zeta}}\right.\\ &\quad\left.+m(\zeta)\left[\frac{A^2+\zeta+\gamma(T(0)-T(x))}{2(A^2+\zeta)}\right.\right.\\ &\quad\left.\left.+\frac{i}{2}\frac{(A^2+\zeta)T(x)+\gamma T(0)T(x)+\gamma\zeta}{(A^2+\zeta)\sqrt{\zeta}}\right]\right\} \end{aligned}$$

which has to belong to $L^2(0,\infty)$. Therefore the coefficient of $\exp[-ix\sqrt{\zeta}]$ must tend to 0 when $x\to\infty$. Using the fact that $T(\infty)=-A$, we arrive at the relation

$$\begin{aligned} &m(\zeta)\{A^2+\zeta+\gamma[T(0)+A]+i\zeta^{-1/2}[-A(A^2+\zeta)-\gamma T(0)A+\gamma\zeta]\}\\ &\quad=T(0)+A+i\zeta^{-1/2}[\zeta-AT(0)], \end{aligned}$$

whence

$$m(\zeta)=\frac{\gamma[\zeta^2+T^2(0)]+T(0)(A^2+\zeta)+i(A^2+\zeta)\sqrt{\zeta}}{[A^2+\zeta+\gamma T(0)]^2+\gamma^2\zeta}.$$

We now observe that the denominator of this expression is always positive on the positive real semi-axis. Therefore, if we take λ to be real, we find from the formula

$$\sigma(\lambda)=\text{const.}+\lim_{\eta\to+0}\mathscr{I}\left\{\frac{1}{\pi}\int_0^\lambda m(\xi+i\eta)\,d\xi\right\}\qquad(\zeta=\xi+i\eta)$$

that, when $\lambda\geqslant 0$,

$$\sigma'(\lambda)=\frac{1}{\pi}\frac{(A^2+\lambda)\sqrt{\lambda}}{[A^2+\lambda+\gamma T(0)]^2+\lambda\gamma^2}.$$

On the semi-axis $\lambda<0$, the function $\sigma(\lambda)$ can have not more than two discontinuities, *i.e.*, the operator cannot have more than two negative eigenvalues. If $\lambda_0<0$ is an eigenvalue, then the magnitude of the jump of $\sigma(\lambda)$ at the point λ_0 is given by

$$\sigma(\lambda_0+\varepsilon)-\sigma(\lambda_0-\varepsilon)$$
$$=\lim_{\eta\to+0}\mathscr{I}\left\{\frac{1}{\pi}\int_{\lambda_0-\varepsilon}^{\lambda_0+\varepsilon}\frac{\gamma[\zeta+T^2(0)]+[T(0)+i\sqrt{\zeta}](A^2+\zeta)}{[A^2+\zeta+\gamma T(0)]^2+\zeta\gamma^2}\,d\xi\right\}.$$

In the general case both the eigenvalues and the magnitude of the jumps of the function $\sigma(\lambda)$ will depend on the parameter B. However, there is one value of γ, namely $\gamma=0$, at which only the jump depends on B, and the (unique) eigenvalue does not depend on B.

Let us complete all the calculations for the case $\gamma=0$. Firstly,

$$\sigma'(\lambda)=\frac{1}{\pi}\frac{\sqrt{\lambda}}{A^2+\lambda}\qquad(\lambda\geqslant 0);\tag{3_1}$$

secondly, $\lambda_0=-A^2$ and

$$\sigma(\lambda_0+0)-\sigma(\lambda_0-0)=A(1+\tanh B),\tag{3_2}$$

because, for sufficiently small $\varepsilon>0$,

$$\sigma(\lambda_0+\varepsilon)-\sigma(\lambda_0-\varepsilon)=\lim_{\eta\to+0}\mathscr{I}\left\{\frac{1}{\pi}\int_{\lambda_0-\varepsilon}^{\lambda_0+\varepsilon}\frac{T(0)+i\sqrt{\zeta}}{A^2+\zeta}\,d\xi\right\}$$
$$=\lim_{\eta\to+0}\mathscr{I}\left\{\frac{1}{\pi}\int_{-A^2-\varepsilon}^{-A^2+\varepsilon}\frac{T(0)-\sqrt{(-\xi-i\eta)}}{(A^2+i\eta)+\xi}\,d\xi\right\}$$
$$=-[T(0)-A]=A(1+\tanh B).$$

The corresponding eigenfunction has the form

$$\varphi_0(x) = \cosh Ax - \tanh(Ax+B)\sinh Ax = \frac{\cosh B}{\cosh(Ax+B)}. \tag{3_3}$$

The boundary condition (2) when $\gamma = 0$ takes the form

$$y'(0) + A\tanh B\,.\,y(0) = 0. \tag{3_4}$$

Thus, under this boundary condition, the spectrum of the equation†

$$-y'' - \frac{2A^2}{\cosh^2(Ax+B)}\,y = \lambda y \tag{1'}$$

does not depend on the parameter B. But, of course, the spectral function does depend on B, for, as the formula (3_2) shows, the magnitude of the jump at one point depends on B.

Another simple case occurs when

$$\gamma T(0) + A^2 = 0$$

i.e., when

$$\gamma = A\coth B \qquad (B \neq 0).$$

In this case σ is continuous and

$$\sigma'(\lambda) = \begin{cases} \dfrac{1}{\pi}\cdot\dfrac{A^2+\lambda}{(\gamma^2+\lambda)\sqrt{\lambda}} & (\lambda > 0) \\ 0 & (\lambda \leqslant 0) \end{cases}$$

and the boundary condition takes the form

$$y'(0) = \frac{A}{\sinh B\cosh B}\,y(0).$$

† If we put $A^2 = c$, $B = -ct\sqrt{c}$, where $c > 0$, then the potential of the equation (1′) takes the form

$$w(x;t) = -\frac{2c}{\cosh^2((x-ct)\sqrt{c})}.$$

If we interpret t here as time, then this function will be a so-called *solitary wave* or *soliton*; it represents a travelling-wave type of solution of a certain non-linear equation, which in our case takes the form

$$w_t = \tfrac{3}{2}ww_x - \tfrac{1}{4}w_{xxx}$$

and differs only in the normalization and choice of units of length and time from the classical equation of Korteweg and de Vries:

$$w_t + ww_x + w_{xxx} = 0.$$

See Lax [69].

We still have to consider the case of the boundary condition (2′). The solution $\psi(x;\lambda)$ of the equation (1) under this boundary condition plays a certain part (as a simple example) in scattering theory.† Namely, in the physical considerations a certain interest attaches to the asymptotic expression for the function $\psi(x;\lambda)$ as $x\to\infty$. In the case of the equation (1) this asymptotic expression is easily found from the formula obtained earlier; it has the form

$$\psi(x;\lambda)\sim N(s)\frac{\sin(xs+\eta(s))}{s}$$

where

$$s=\sqrt{\lambda}>0,\qquad N(s)=\left|\frac{s-iA\tanh B}{s-iA}\right|=\sqrt{\left(\frac{s^2+A^2\tanh^2 B}{s^2+A^2}\right)}$$

and

$$\eta(s)=\arg\frac{s-iA\tanh B}{s-iA}=\tan^{-1}\frac{As\exp[-B]}{s^2\cosh B+A^2\sinh B}.$$

$N(s)$ is called the *limiting amplitude* and $\eta(s)$ the *limiting phase difference.*

The procedure for finding the spectral function $\tau(\lambda)$ corresponding to the condition (2′) is similar to that used for the condition (2). The final formula has the form

$$\tau'(\lambda)=\begin{cases}\dfrac{1}{\pi}\dfrac{(A^2+\lambda)\sqrt{\lambda}}{\lambda+A^2\tanh^2 B} & (\lambda\geq 0),\\ 0 & (\lambda<0).\end{cases}$$

2. *The classical Fourier integral* on the half-line and its generalization are obtained from the preceding example *1* by letting $B\to\pm\infty$. So we can restrict ourselves to a formulation of the result.

Thus, let us consider the differential expression

$$l=-\frac{d^2}{dt^2}\qquad (0\leq t<\infty).$$

The solution of the equation

$$l[y]=\lambda y$$

under the boundary condition‡

$$y'(0)=\mp Ay(0) \tag{4}$$

† See R. G. Newton [96]; Newton considers the particular case of an Ekkart potential. Our notation differs from that adopted in scattering theory.

‡ Here, and subsequently, when we write two signs, the upper one refers to the case $B=\infty$, and the lower one to the case $B=-\infty$, in the corresponding formulae of example *1* (in all cases, $A\geq 0$).

is the function

$$\varphi(t;\lambda)=\cos(t\sqrt{\lambda})\mp A\frac{\sin(t\sqrt{\lambda})}{\sqrt{\lambda}}.$$

Under the boundary condition (4) the inversion formulae have the form:

(a) in the case of the upper sign and $A>0$:

$$\begin{cases}\Phi(\lambda)=\displaystyle\int_0^\infty f(t)\left[\cos(t\sqrt{\lambda})-A\frac{\sin(t\sqrt{\lambda})}{\sqrt{\lambda}}\right]dt & (\lambda\geqslant 0),\\ \Phi(-A^2)=\displaystyle\int_0^\infty f(t)e^{-At}\,dt,\end{cases}$$

$$f(t)=2A\Phi(-A^2)e^{-At}+\frac{1}{\pi}\int_0^\infty \Phi(\lambda)\left[\cos(t\sqrt{\lambda})-A\frac{\sin(t\sqrt{\lambda})}{\sqrt{\lambda}}\right]\frac{\sqrt{\lambda}}{A^2+\lambda}\,d\lambda;$$

(b) in the case of the lower sign and $A\geqslant 0$:

$$\Phi(\lambda)=\int_0^\infty f(t)\left[\cos(t\sqrt{\lambda})+A\frac{\sin(t\sqrt{\lambda})}{\sqrt{\lambda}}\right]dt,$$

$$f(\lambda)=\frac{1}{\pi}\int_0^\infty \Phi(\lambda)\left[\cos(t\sqrt{\lambda})+A\frac{\sin(t\sqrt{\lambda})}{\sqrt{\lambda}}\right]\frac{\sqrt{\lambda}}{A^2+\lambda}\,d\lambda.$$

Under the boundary condition

$$y(0)=0 \tag{4'}$$

the solution is

$$\psi(t;\lambda)=\frac{\sin(t\sqrt{\lambda})}{\sqrt{\lambda}}$$

and the inversion formulae take the form

$$\Phi(\lambda)=\int_0^\infty f(t)\frac{\sin(t\sqrt{\lambda})}{\sqrt{\lambda}}\,dt,\qquad f(t)=\frac{1}{\pi}\int_0^\infty \Phi(\lambda)\frac{\sin(t\sqrt{\lambda})}{\sqrt{\lambda}}\sqrt{\lambda}\,d\lambda.$$

3. The Legendre functions. The Legendre differential expression has the form

$$l=-\frac{d}{dt}(1-t^2)\frac{d}{dt}.$$

When discussing the Legendre equation

$$l[u]-\lambda u=0,$$

it is convenient to express the parameter λ in the form $\lambda=\mu(\mu+1)$. If we make the change of variable

$$x=(1-t)/2,$$

then the Legendre equation takes the form

$$x(1-x)\frac{d^2u}{dx^2}+(1-2x)\frac{du}{dx}+\mu(\mu+1)u=0.$$

This is a particular case of the hypergeometric equation

$$x(1-x)\frac{d^2u}{dx^2}+\{C-(a+b+1)x\}\frac{du}{dx}-abu=0.$$

Therefore one of the solutions of the Legendre equation is the hypergeometric function

$$P_\mu(t)=F(\mu+1;-\mu;1;(1-t)/2)=1+$$

$$+\sum_{k=1}^{\infty}\frac{(\mu+1)(\mu+2)\cdots(\mu+k)(-\mu)(-\mu+1)\cdots(-\mu+k-1)}{(k!)^2}\left(\frac{1-t}{2}\right)^k.$$

For subsequent use we note that

$$P'_\mu(t)=\frac{\mu(\mu+1)}{2}F(\mu+2;-\mu+1;2;(1-t)/2).$$

The series for $P_\mu(t)$ will terminate, *i.e.*, will become a polynomial (a *Legendre polynomial*) if and only if μ is an integer. In general, however, $P_\mu(t)$ is called a *Legendre function of the first kind of order* μ. The region of convergence of the series for $P_\mu(t)$ is the disc

$$|t-1|<2.$$

Another solution of the Legendre equation is the function $P_\mu(-t)$. Both the functions $P_\mu(t)$ and $P_\mu(-t)$ are regular in the disc $|t|<1$.

From the general formulae†

$$\lim_{x\to1-0}\frac{F(a;b;a+b;x)}{\ln(1/(1-x))}=\frac{\Gamma(a+b)}{\Gamma(a)\Gamma(b)},$$

$$\lim_{x\to1-0}(1-x)^{a+b-c}F(a;b;c;x)=\frac{\Gamma(c)\Gamma(a+b-c)}{\Gamma(a)\Gamma(b)},$$

$$\mathcal{R}(a+b-c)>0,$$

† See Whittaker and Watson [127] or Bateman [12].

the following relations hold for the Legendre functions:

$$\lim_{t\to-1+0} \frac{P_\mu(t)}{\ln((1-t)/(1+t))} = -\frac{\sin \pi\mu}{\pi},$$

$$\lim_{t\to-1+0} P_\mu^{[1]}(t) = \lim_{t\to-1+0} (1-t^2)P'_\mu(t) = \frac{2\sin \pi\mu}{\pi}.$$

From the first of these relations and the equality $P_\mu(1)=1$ we conclude that, when μ is not an integer, the functions $P_\mu(t)$ and $P_\mu(-t)$ are linearly independent. Let us assume that μ is not an integer; it may be real or non-real.

We shall consider the differential expression l first on the interval $(-1, 1)$ and then on the half-line $(1, \infty)$.

a) *The case of the interval* $(-1, 1)$. In this case both the functions $P_\mu(t)$ and $P_\mu(-t)$ belong to $L^2(-1, 1)$. Moreover, the quasi-derivatives $P_\mu^{[1]}(t)$ and $P_\mu^{[1]}(-t)$ are absolutely continuous, and the quasi-derivatives of second order belong to $L^2(-1, 1)$ by virtue of the Legendre differential equation. Therefore the equation

$$(L^*-\lambda I)u = 0 \qquad (-1<t<1), \tag{5}$$

when μ is not an integer, has two linearly independent solutions, and so the deficiency index of the operator L is $(2, 2)$.

We recommend the reader to verify also that the orthonormal system of solutions of the equation (5) has the form

$$u_1(t;\lambda) = A_\mu\{P_\mu(t)+P_\mu(-t)\},$$
$$u_2(t;\lambda) = B_\mu\{P_\mu(t)-P_\mu(-t)\},$$

where A_μ, B_μ are constants whose arguments are arbitrary. We shall therefore take $A_\mu>0$, $B_\mu>0$.

We now put

$$\omega^*(t) = \omega(t) - \tfrac{1}{2}\omega^{[1]}(t)\ln\frac{1+t}{1-t},$$

where the quasi-derivative $\omega^{[1]}(t)$ is equal to $(1-t^2)\omega'(t)$. We write the Lagrange bilinear form as

$$[\varphi, \psi](t) = \varphi^*(t)\overline{\psi^{[1]}(t)} - \varphi^{[1]}(t)\overline{\psi^*(t)}.$$

It is easy to show that:

1) for every function $\varphi \in D(L^*)$ the limits

$$\lim_{t\to\pm1} \varphi^{[1]}(t) \qquad \text{and} \qquad \lim_{t\to\pm1} \varphi^*(t)$$

exist and are finite;

2) the manifold $D(L)$ is the set of all those $\varphi(t) \in D(L^*)$ for which these four limits are equal to 0.

The proof follows from the fact that, for any function $f \in L^2(-1, 1)$, the general integral of the equation $l[y] = f(t)$ has the form

$$y(t) = \left\{C_1 + \tfrac{1}{2}\int_{-1}^{t} f(s) \ln \frac{1+s}{1-s}\, ds\right\} + \left\{C_2 - \tfrac{1}{2}\int_{-1}^{t} f(s)\, ds\right\} \ln \frac{1+t}{1-t},$$

where C_1, C_2 are constants, and, because of this representation,

$$y^{[1]}(t) = 2\left\{C_2 - \tfrac{1}{2}\int_{-1}^{t} f(s)\, ds\right\}, \qquad y^*(t) = C_1 + \tfrac{1}{2}\int_{-1}^{t} f(s) \ln \frac{1+s}{1-s}\, ds.$$

In order to obtain a self-adjoint extension of the operator L, we have, in accordance with the general theory, to fix on some value of λ in the upper half-plane (we shall take $\lambda = \mu(\mu+1)$, where μ is a pure imaginary) and, having chosen a unitary matrix $[\theta_{ik}]_{i,k=1}^{2}$, we put

$$w_1(t) = w_1(t; \lambda) = u_1(t; \lambda) + \theta_{11}\overline{u_1(t; \lambda)} + \theta_{12}\overline{u_2(t; \lambda)},$$
$$w_2(t) = w_2(t; \lambda) = u_2(t; \lambda) + \theta_{21}\overline{u_1(t; \lambda)} + \theta_{22}\overline{u_2(t; \lambda)}.$$

Further, we have to find the values of

$$w_k^{[1]}(\pm 1), \qquad w_k^*(\pm 1) \qquad (k = 1, 2)$$

and, by means of them, to set up the boundary conditions

$$[\varphi, w_k]_{-1}^{1} = 0 \qquad (k = 1, 2),$$

separating the domain $D(L^*)$ from the domain $D(L_\theta)$.

Using the given formulae it is not difficult to establish the following table:

$$w_1^{[1]}(\pm 1) = \frac{2 \sin \pi\mu}{\pi}\{\mp A_\mu(1-\theta_{11}) - B_\mu\theta_{12}\},$$

$$w_2^{[1]}(\pm 1) = \frac{2 \sin \pi\mu}{\pi}\{\pm A_\mu\theta_{21} + B_\mu(1-\theta_{22})\},$$

$$w_1^*(\pm 1) = A_\mu\{[1+\gamma(\mu)] + \theta_{11}[1+\overline{\gamma(\mu)}]\} \pm \theta_{12}B_\mu[1-\overline{\gamma(\mu)}],$$

$$w_2^*(\pm 1) = \theta_{21}A_\mu[1+\overline{\gamma(\mu)}] \pm B_\mu\{[1-\gamma(\mu)] + \theta_{22}[1-\overline{\gamma(\mu)}]\}.$$

We shall not go into the calculation of the number $\gamma(\mu)$, but it will become clear shortly that $\gamma(\mu) \neq \pm 1$.

We obtain the simplest and most important self-adjoint extension by putting

$$\theta_{11} = \theta_{22} = 1, \qquad \theta_{12} = \theta_{21} = 0.$$

In this case,

$$w_1^{[1]}(\pm 1) = 0, \qquad w_2^{[1]}(\pm 1) = 0,$$

and

$$w_1^*(\pm 1) = A_\mu[2 + \gamma(\mu) + \overline{\gamma(\mu)}],$$
$$w_2^*(\pm 1) = \pm B_\mu[2 - \gamma(\mu) - \overline{\gamma(\mu)}].$$

Since neither $w_1(t)$ nor $w_2(t)$ belongs to $D(L)$, the last expression is not zero, and therefore $\gamma(\mu) \neq \pm 1$.

The boundary conditions characterizing this extension take the form

$$\varphi^{[1]}(1) = \varphi^{[1]}(-1) = 0. \tag{A}$$

We can put these boundary conditions into a different form. Firstly, they are equivalent to the condition

$$\frac{1}{\sqrt{(1-t^2)}}\,\varphi^{[1]}(t) \in L^2(-1, 1). \tag{B}$$

Secondly, they are equivalent to the requirement that:

$$\varphi(t) \text{ shall tend to finite limits as } t \to \pm 1. \tag{C}$$

We shall prove that these conditions (A), (B), (C) are equivalent; in doing so, it suffices to consider the case of a real-valued function $\varphi(t)$.

We write down the identity

$$-\int_\alpha^\beta \varphi(t)[(1-t^2)\varphi'(t)]'\,dt = [-(1-t^2)\varphi'(t)\varphi(t)]_\alpha^\beta$$
$$+\int_\alpha^\beta (1-t^2)\varphi'^2(t)\,dt, \qquad (-1<\alpha<\beta<1) \tag{6}$$

the left-hand side of which has a finite limit when $\alpha \to -1$ and $\beta \to +1$ whatever the function $\varphi(t) \in D(L^*)$ may be. On the other hand.

$$\varphi^{[1]}(t) = \varphi^{[1]}(-1) - \int_{-1}^t l[\varphi(s)]\,ds = \varphi(1) + \int_t^1 l[\varphi(s)]\,ds.$$

Therefore, if the condition (A) holds, then

$$|\varphi^{[1]}(t)| \leqslant \sqrt{(1 \pm t)}\,.\,\|L^*\varphi\|,$$

and so

$$\lim_{t \to \pm 1} \varphi^{[1]}(t) \ln \frac{1+t}{1-t} = 0.$$

Since $\varphi^*(t)$ tends to finite limits when $t \to \pm 1$, the function $\varphi(t)$ will tend to the same limits, *i.e.*, the condition (C) holds. Thus (A) implies (C).

Now suppose the condition (C) holds. Since, by the proposition 1) on p. 537, $\varphi^{[1]}(t)$ has finite limits as $t \to \pm 1$, the first term on the right in (6) has a finite limit as $\alpha \to -1$ and $\beta \to +1$ and the left-hand side of (6) has the same property. Therefore the second term on the right has a finite limit†

$$J[\varphi] = \int_{-1}^{1} (1-t^2)\varphi'^2(t)\,dt < \infty \tag{7}$$

as $\alpha \to -1$ and $\beta \to +1$. That is to say, the condition (B) holds. Thus (C) implies (B).

Finally, suppose that the condition (B) holds, *i.e.*, the relation (7) holds. Then $\varphi^{[1]}(1)$ cannot be different from 0, since otherwise the fact that $\varphi^*(1)$ is finite would imply that $\lim_{t\to 1} \varphi(t)$ is infinite, and, by (6), this leads to a contradiction. Similarly one proves that $\varphi^{[1]}(-1) = 0$. Thus (B) implies (A), and our assertion has been proved.

The self-adjoint extension characterized by one of the conditions (A), (B), or (C) leads to the classical expansion in Legendre polynomials. Since the boundary conditions (A) are separated, the spectrum of the extension is simple. Moreover, it is purely a point spectrum without any finite points of condensation, and the inverse operator is completely continuous. The Legendre polynomials

$$P_n(t) = \frac{1}{2^n n!}\frac{d^n(t^2-1)^n}{dt^n} \qquad (n = 0, 1, 2, \ldots)$$

satisfy the differential equation

$$-\frac{d}{dt}(1-t^2)\frac{du}{dt} - n(n+1)u = 0$$

and also the boundary condition (C). And since the set of all the polynomials $P_n(t)$ forms a complete orthogonal system in $L^2(-1, 1)$, the spectrum of the extension under consideration consists of the points $n(n+1)$ $(n = 0, 1, 2, \ldots)$, and the inversion formulae in this case follow from the expansion as a series of Legendre polynomials.

b) *The case of the half-line* $(1, \infty)$

We now consider the equation,

$$-\frac{d}{dt}\left((1-t^2)\frac{du}{dt}\right) - \mu(\mu+1)u = 0$$

on the half-line $t > 1$. There are relations for the hypergeometric function

† We observe that the equation $l[\varphi] = 0$ is the Euler–Lagrange equation for the integral $J[\varphi]$. In this connection, see K. O. Friedrichs [30].

which enable it to be continued analytically from the unit disc on to other regions of the complex plane. Thanks to these relations, the following representation† holds, when $t>1$, for the Legendre functions of the first kind:

$$P_\mu(t)=\frac{1}{\sqrt{\pi}}\cdot\frac{1}{2^{\mu+1}}\cdot\frac{\Gamma(-\frac{1}{2}-\mu)}{\Gamma(-\mu)}\cdot\frac{1}{t^{\mu+1}}\cdot F\left(1+\frac{\mu}{2};\frac{1+\mu}{2};\frac{3}{2}+\mu;\frac{1}{t^2}\right)$$
$$+\frac{1}{\sqrt{\pi}}\cdot 2^\mu\cdot\frac{\Gamma(\frac{1}{2}+\mu)}{\Gamma(1+\mu)}\cdot t^\mu\cdot F\left(\frac{1-\mu}{2};-\frac{\mu}{2};\frac{1}{2}-\mu;\frac{1}{t^2}\right). \tag{8}$$

We take as the second solution of the Legendre equation on the half-line $t>1$ the *Legendre function of the second kind*‡

$$Q_\mu(t)=\frac{\sqrt{\pi}}{2^{\mu+1}}\cdot\frac{\Gamma(1+\mu)}{\Gamma(\frac{3}{2}+\mu)}\cdot\frac{1}{t^{\mu+1}}\cdot F\left(1+\frac{\mu}{2};\frac{1+\mu}{2};\frac{3}{2}+\mu;\frac{1}{t^2}\right). \tag{9}$$

We now put $\mu(\mu+1)=-\frac{1}{4}-\lambda$, where λ is a new parameter. Thus $\mu=-\frac{1}{2}+i\sqrt{\lambda}$ and then the Legendre equation can be written in the form

$$-\frac{d}{dt}(t^2-1)\frac{du}{dt}-\tfrac{1}{4}u=\lambda u.$$

In our inversion formulae the parameter λ will be $\geqslant 0$.

The series written above for $P_\mu(t)$ converges outside the unit disc in the t-plane, and the expansion considered earlier converges in the disc $|t-1|<2$. Therefore the point $t=1$ is a regular point for the function $P_\mu(t)$, and, using the expansion in the disc $|t-1|<2$, we can write down at once that

$$P_\mu(1)=1, \qquad [(t^2-1)P_\mu'(t)]_{t=1}=0.$$

We also see that the function $P_{-\frac{1}{2}+i\sqrt{\lambda}}(t)$ is real when $t>1$ and $\lambda\geqslant 0$ is real.

Together with $Q_\mu(t)$, the function $Q_{-1-\mu}(t)$ will also be a solution of the differential equation, and so, of course, $Q_{-1-\mu}(t)$ can be expressed as a linear combination of $P_\mu(t)$ and $Q_\mu(t)$. As the second solution of the Legendre equation we associate with the function $P_{-\frac{1}{2}+i\sqrt{\lambda}}(t)$ the function

$$\psi(t;\lambda)=\tfrac{1}{2}\{Q_{-\frac{1}{2}+i\sqrt{\lambda}}(t)+Q_{-\frac{1}{2}-i\sqrt{\lambda}}(t)\},$$

which, clearly, is real on the half-line $t>1$ when $\lambda\geqslant 0$. By the general properties mentioned earlier of the hypergeometric function, and using

† See Bateman [12], Chap. III.3.2.
‡ Here we have used the formula

$$\Gamma(2z)=\frac{1}{\sqrt{\pi}}2^{2z-1}\Gamma(z)\Gamma(z+\tfrac{1}{2}).$$

again the formula in the second footnote on p. 541, we obtain the following relations for $\psi(t;\lambda)$:

$$\lim_{t\to 1+0}\frac{\psi(t;\lambda)}{\ln(1-1/t^2)}=-\tfrac{1}{2},\qquad \lim_{t\to 1+0}(t^2-1)\psi'(t;\lambda)=-1.$$

Hence we conclude that

$$(t^2-1)\begin{vmatrix} P_{-\frac{1}{2}+i\sqrt{\lambda}}(t) & P'_{-\frac{1}{2}+i\sqrt{\lambda}}(t)\\ \psi(t;\lambda) & \psi'(t;\lambda)\end{vmatrix}=-1.$$

Let us find the deficiency index of the operator L. If $1<c<a$, then in the interval $(1, c]$ with one singular end-point the deficiency index is (2, 2) because both solutions belong to $L^2(1, c)$; further, in the interval $[c, \infty)$, the deficiency index of the operator L is (1, 1), and therefore, by Theorem 3 in §127, the deficiency index in $(1, \infty)$ is (1, 1).

We now construct the inversion formulae for that extension of the operator L which is determined by the boundary condition

$$[(t^2-1)u'(t)]_{t=1+0}=0.$$

The function $\varphi(t;\lambda)$ which is equal to 1 when $t=1$ will be $P_{-\frac{1}{2}+i\sqrt{\lambda}}(t)$. To obtain the inversion formula in this case we can use the method of direction functionals described in §129, if we observe that, in constructing the Green's function, it is not essential that there should be a solution $v(t;\lambda)$ which is equal to 0 at the left-hand end-point (here, where $t=1$). In our case we can take as the function $v(t;\lambda)$ the function $\psi(t;\lambda)$; this has a logarithmic singularity at the left-hand end-point, but all the considerations of §129, including the final formula for the spectral function $\sigma(\lambda)$, go through. We shall not go into the details of the simple proof of this assertion, but pass on instead to the construction, proceeding in the same way as in the case where the left-hand end-point is regular. Thus, assuming that $0<\arg\zeta<\pi$, we try to determine a function $m(\zeta)$ $(\mathscr{I}(\zeta)>0)$ from the condition that the linear combination

$$\psi(t;\lambda)+m(\zeta)P_{-\frac{1}{2}+i\sqrt{\zeta}}(t) \tag{10}$$

shall belong to $L^2(1, \infty)$. From the form of the expansions of (8) and (9) it is easily seen that $m(\zeta)$ is uniquely determined. For, as $t\to\infty$, the hypergeometric function tends to the value 1, and the first of the two factors $t^{-1-\mu}$ and t^{μ} (when $\mu=-\frac{1}{2}+i\sqrt{\zeta}$ and $\mathscr{I}(\zeta)>0$) does not belong to $L^2(1, \infty)$, but the second one does. Thus we obtain the equation

$$\frac{1}{2}\cdot\frac{\sqrt{\pi}}{2^{\mu+1}}\cdot\frac{\Gamma(1+\mu)}{\Gamma(\frac{3}{2}+\mu)}+m(\zeta)\cdot\frac{1}{\sqrt{\pi}}\cdot\frac{1}{2^{\mu+1}}\cdot\frac{\Gamma(-\frac{1}{2}-\mu)}{\Gamma(-\mu)}=0.$$

Solving it, we find that

$$m(\zeta)=-\frac{\pi}{2}\cdot\frac{\Gamma(1+\mu)\Gamma(-\mu)}{\Gamma(\frac{3}{2}+\mu)\Gamma(-\frac{1}{2}-\mu)}=-\frac{\pi}{2}\cdot\frac{\sin(\frac{1}{2}+\mu)\pi}{\sin\pi\mu}=\frac{\pi i}{2}\cdot\tanh(\pi\sqrt{\zeta}).$$

We see that the function of the form (10) which belongs to $L^2(1,\infty)$ is equal to†

$$\tfrac{1}{2}\{Q_{-\frac{1}{2}+i\sqrt{\zeta}}(t)+\pi i\tanh(\pi\sqrt{\zeta})\,.\,P_{-\frac{1}{2}+i\sqrt{\zeta}}(t)\}. \tag{10'}$$

It is now easy to write down the inversion formulae

$$g(\lambda)=\int_1^\infty f(t)P_{-\frac{1}{2}+i\sqrt{\lambda}}(t)\,dt,$$

$$f(t)=\tfrac{1}{2}\int_1^\infty g(\lambda)P_{-\frac{1}{2}+i\sqrt{\lambda}}(t)\,.\,\tanh(\pi\sqrt{\lambda})\,d\lambda.$$

These formulae are known as the *Mehler–Fock formulae.*

The Parseval equation here is

$$\int_1^\infty |f(t)|^2\,dt=\tfrac{1}{2}\int_0^\infty |g(\lambda)|^2\tanh(\pi\sqrt{\lambda})\,d\lambda.$$

4. The Chebyshev–Hermite functions. These functions are connected with the differential expression

$$l=-\frac{d^2}{dt^2}+t^2 \qquad (-\infty<t<\infty) \tag{11}$$

which is of great importance in quantum mechanics, where it appears in the theory of the 'linear oscillator'.

The operator L generated by the differential expression (11) is self-adjoint, by Theorem 5 in §128. It is easily verified that the equation

$$-u''+(t^2-\lambda)u=0$$

with $\lambda=2n+1$ $(n=0,1,2,\ldots)$ is satisfied by the successive Chebyshev–Hermite polynomials (see §12), which form a complete system in $L^2(-\infty,\infty)$. Therefore the operator L has a purely discrete system, with a unique limit-point at infinity. Thus the inversion formulae associated with the operator L reduce to the expansion in terms of the Chebyshev–Hermite functions.

† Here it should be pointed out that the following formula (mentioned earlier) is proved in the theory of Legendre functions:

$$Q_{-\frac{1}{2}-i\sqrt{\zeta}}(t)-Q_{-\frac{1}{2}+i\sqrt{\zeta}}(t)=\pi i\tanh(\pi\sqrt{\zeta})\,.\,P_{-\frac{1}{2}+i\sqrt{\zeta}}(t),$$

and by using this we can see immediately that (10′) belongs to $L^2(1,\infty)$ when $\mathscr{I}(\sqrt{\zeta})>0$.

This example is, in some respects, particularly instructive. It shows that the resolvent's property of being a completely continuous operator —— a property which, by Theorem 2 of §128 always holds in the quasi-regular case —— may also hold in other cases (even when, as in this example, the deficiency numbers are minimal).

5. *Bessel functions.* Of the various differential expressions which lead to Bessel functions the most important one is of the form

$$l = -\frac{d^2}{dt^2} + \frac{\nu^2 - 1/4}{t^2}, \tag{12}$$

and it is this differential expression, with the parameter $\nu \geqslant 0$, which we shall consider. As regards the interval, it is natural to investigate the following three cases:

α) $0 < t \leqslant 1$ —— one singular end-point,
β) $1 \leqslant t < \infty$ —— one singular end-point,
γ) $0 < t < \infty$ —— two singular end-points.

The general integral of the equation

$$l[u] - \lambda u = 0 \tag{13}$$

for $\lambda \neq 0$ has the form

$$u(t;\lambda) = At^{1/2} J_\nu(t\sqrt{\lambda}) + Bt^{1/2} Y_\nu(t\sqrt{\lambda}), \tag{14}$$

where A, B are arbitrary constants, and $J_\nu(z)$ and $Y_\nu(z)$ are the cylinder functions of the first and second kind respectively, defined by the formulae

$$J_\nu(z) = \sum_{k=0}^{\infty} \frac{(-1)^k (z/2)^{\nu+2k}}{k!\,\Gamma(\nu+k+1)}, \qquad Y_\nu(z) = \frac{J_\nu(z)\cos \pi\nu - J_{-\nu}(z)}{\sin \pi\nu}.$$

We shall assume that $\mathscr{I}(\lambda) > 0$ and $t > 0$; thus $z = t\sqrt{\lambda}$ will satisfy the inequality $0 < \arg z < \pi/2$. But if $|\arg z| < \pi - \varepsilon$ and if $|z| \to \infty$, then the following asymptotic formulae will hold:

$$J_\nu(z) + iY_\nu(z) \equiv H_\nu^{(1)}(z) \sim \sqrt{\left(\frac{2}{\pi z}\right)} \exp\left[i\left(z - \frac{\pi\nu}{2} - \frac{\pi}{4}\right)\right],$$

$$J_\nu(z) - iY_\nu(z) \equiv H_\nu^{(2)}(z) \sim \sqrt{\left(\frac{2}{\pi z}\right)} \exp\left[-i\left(z - \frac{\pi\nu}{2} - \frac{\pi}{4}\right)\right].$$

By means of the formula (14) we find that the equation (13) does not have two linearly independent solutions belonging to $L^2(1, \infty)$. But one such solution necessarily exists; it is the function

$$u_1(t;\lambda)=At^{1/2}H_\nu^{(1)}(t\surd\lambda).$$

In the case β), then, the deficiency index of the operator L is (1, 1).

Let us find the deficiency index of L in the case α). Since

$$z^{1/2}J_\nu(z)\sim\frac{z^{\nu+1/2}}{2^\nu\Gamma(\nu+1)}\qquad\text{when } z\to 0,$$

and, on the other hand, $z^{1/2}Y_\nu(z)$ does not belong to $L^2(0, 1)$ if $\nu\geq 1$, but does belong to $L^2(0, 1)$ if $0\leq\nu<1$, it follows that, in the case α), the deficiency index of L is

$$\begin{aligned}&(2,2)\quad\text{if}\quad 0\leq\nu<1,\\ &(1,1)\quad\text{if}\quad \nu\geq 1.\end{aligned}$$

Therefore, in the case γ), the deficiency index of L is

$$\begin{aligned}&(1,1)\quad\text{if}\quad 0\leq\nu<1,\\ &(0,0)\quad\text{if}\quad \nu\geq 1.\end{aligned}$$

In each of the cases discussed, the differential expression (12) generates certain inversion formulae.

We obtain a unique pair of formulae for the interval $(0,\infty)$ when $\nu\geq 1$. This pair has the form

$$\left.\begin{aligned}g(\lambda)&=\int_0^\infty\surd(\lambda t)\,.\,J_\nu(\lambda t)f(t)\,dt,\\ f(t)&=\int_0^\infty\surd(\lambda t)\,.\,J_\nu(\lambda t)g(\lambda)\,d\lambda.\end{aligned}\right\}\tag{15}$$

We have here not just a unitary, but a self-adjoint, operator in $L^2(0,\infty)$; it is called the *Hankel transformation.* When $0\leq\nu<1$ the formulae (15) are still true, but they are no longer the unique inversion formulae† generated by the expression (12) on the half-line.

The validity of the inversion formulae (15) is best verified directly (this will be done when discussing example 6 below) by using one of the methods we have indicated for obtaining Fourier–Plancherel inversion formulae. We shall see that the formula (15) is valid even when $\nu>-1$ (and therefore for $\nu\geq -1$, since $J_{-1}(z)=-J_1(z)$).

Let us consider the case of the interval (0, 1) with $\nu\geq +1$, and suppose that the boundary condition at the regular end-point has the form

† See Titchmarsh [124], vol. I, §4.11, for the construction of all the expansions relating to the case $0\leq\nu<1$.

$u(1)=0$. We can therefore put

$$u(t;\lambda)=\frac{\pi}{2}\sqrt{t}\,.\,\{J_\nu(t\sqrt{\lambda})Y_\nu(\sqrt{\lambda})-Y_\nu(t\sqrt{\lambda})J_\nu(\sqrt{\lambda})\};$$

also let

$$v(t;\lambda)=\frac{\pi}{2}\sqrt{(t\lambda)}\,.\,\{J_\nu(t\sqrt{\lambda})Y'_\nu(\sqrt{\lambda})-Y_\nu(t\sqrt{\lambda})J'_\nu(\sqrt{\lambda})\},$$

so that, using the formula

$$J_\nu(z)Y'_\nu(z)-Y_\nu(z)J'_\nu(z)=2/(\pi z),$$

we have

$$u(1;\lambda)=0, \qquad u'(1;\lambda)=-1,$$
$$v(1;\lambda)=0, \qquad v'(1;\lambda)=1/2.$$

Then

$$K(t,s;\zeta)=\begin{cases}u(t;\zeta)[v(s;\zeta)+m(\zeta)u(s;\zeta)] & (t\geqslant s),\\ u(s;\zeta)[v(t;\zeta)+m(\zeta)u(t;\zeta)] & (t\leqslant s),\end{cases}$$

where the function $m(\zeta)$ is determined from the fact that the solution

$$v(t;\zeta)+m(\zeta)u(t;\zeta) \qquad (\mathscr{I}(\zeta)>0)$$

belongs to the space $L^2(0,1)$. This implies that the function

$$\frac{\pi}{2}\sqrt{t}\,.\,J_\nu(t\sqrt{\zeta})[\sqrt{\zeta}Y'_\nu(\sqrt{\zeta})+m(\zeta)Y_\nu(\sqrt{\zeta})]$$
$$-\frac{\pi}{2}\sqrt{t}\,.\,Y_\nu(t\sqrt{\zeta})[\sqrt{\zeta}J'_\nu(\sqrt{\zeta})+m(\zeta)J_\nu(\sqrt{\zeta})]$$

must converge to its first member, and so

$$m(\zeta)=-\sqrt{\zeta}\,.\,\frac{J'_\nu(\sqrt{\zeta})}{J_\nu(\sqrt{\zeta})}.$$

This function can be expressed in the form

$$m(\zeta)=2\sum_{n=1}^{\infty}\frac{\zeta}{\lambda_n-\zeta}-\nu,$$

where the λ_n are the non-zero roots of the Bessel function $J_\nu(\sqrt{\zeta})$. Hence it follows that $\sigma(\lambda)$ is a step-function, and that the sequence $\lambda_1, \lambda_2, \lambda_3, \ldots$ constitutes the spectrum. The eigenfunctions are

$$u_n(t)=u_n(t;\lambda_n)=-\frac{\sqrt{t}\,.\,J_\nu(t\sqrt{\lambda_n})}{\sqrt{\lambda_n}\,.\,J'_\nu(\sqrt{\lambda_n})},$$

and the inversion formulae reduce to the expansion formula

$$f(t)=\sum_{n=1}^{\infty} 2\lambda_n u_n(t;\lambda)\int_0^1 f(s)u_n(s;\lambda_n)\,ds,$$

known as the *Fourier–Bessel formula.*

If we had taken the more general condition

$$u'(1)=-\theta u(1)$$

at the regular end-point, then we would have obtained the *Fourier–Dini series.*

We shall not discuss the case of the interval (0, 1] when $0\leqslant\nu<1$, but merely mention that the method used for the Legendre differential expression can also be used here.

We recommend the reader to consider the case β) with the interval $[1,\infty)$ and the deficiency index (1, 1). If the boundary condition at the regular end-point is taken to be

$$u(1)=0,$$

then, using the method repeatedly employed, we find that

$$m(\zeta)=-\sqrt{\zeta}\,.\frac{J_\nu'(\sqrt{\zeta})+iY_\nu'(\sqrt{\zeta})}{J_\nu(\sqrt{\zeta})+iY_\nu(\sqrt{\zeta})},\qquad (\mathscr{I}(\zeta)>0)$$

and we obtain the *Weber–Titchmarsh inversion formulae*

$$\Phi(\lambda)=\int_1^{\infty}\sqrt{t}\,.\{J_\nu(t\sqrt{\lambda})Y_\nu(\sqrt{\lambda})-Y_\nu(t\sqrt{\lambda})J_\nu(\sqrt{\lambda})\}f(t)\,dt,$$

$$f(t)=\int_0^{\infty}\sqrt{t}\cdot\frac{J_\nu(t\sqrt{\lambda})Y_\nu(\sqrt{\lambda})-Y_\nu(t\sqrt{\lambda})J_\nu(\sqrt{\lambda})}{2\{J_\nu^2(\sqrt{\lambda})+Y_\nu^2(\sqrt{\lambda})\}}\,\Phi(\lambda)\,d\lambda.$$

6. *The Hankel transformation and the Laguerre–Sonine polynomials*

We shall examine the Bessel function $J_\nu(t)$ $(t>0)$ when $\nu>-1$. It has as a majorant the *modified Bessel function* (of order ν)

$$I_\nu(t)=\sum_{k=0}^{\infty}\frac{(t/2)^{\nu+2k}}{k!\,\Gamma(\nu+k+1)},$$

for which, by using the simple inequality

$$k!\,\Gamma(\nu+k+1)>k!\,(k-1)!\geqslant\frac{(2k-2)!}{2^{2k-2}}\qquad (\nu>-1;\,k=1,2,3,\ldots),$$

we obtain the following (crude) estimate

$$I_\nu(t) \leqslant \frac{1}{\Gamma(\nu+1)}\left(\frac{t}{2}\right)^\nu + \left(\frac{t}{2}\right)^{\nu+2} \cosh t.$$

We shall need the identity

$$\int_0^\infty e^{-r^2/2} r^{\nu+1/2} \sqrt{(rt)} \,.\, J_\nu(rt)\, dr = e^{-t^2/2} t^{\nu+1/2},$$

which is obtained by termwise integration of the series expansion of $J_\nu(rt)$, this being admissible thanks to the above estimate. This identity shows that the Hankel operator

$$(Hf)(t) = \int_0^\infty f(r)\sqrt{(rt)} \,.\, J_\nu(rt)\, dr$$

has $e^{-t^2/2} t^{\nu+1/2}$ as the eigenfunction corresponding to the eigenvalue 1. If we put $r = \alpha y$, $t = \beta x$, where $\alpha > 0$ and $\beta = 1/\alpha$, then the above identity takes the form

$$\int_0^\infty \exp[-\alpha^2 y^2/2] \alpha^{\nu+1} y^{\nu+1/2} \sqrt{(xy)} \,.\, J_\nu(xy)\, dy$$

$$= \exp[-\beta^2 x^2/2] \beta^{\nu+1} x^{\nu+1/2}.$$

Noting that

$$\alpha \frac{\partial}{\partial \alpha} = -\beta \frac{\partial}{\partial \beta},$$

we apply the operator $(\alpha(\partial/\partial\alpha))^n$ to the left-hand side of the last identity, and $(-1)^n(\beta(\partial/\partial\beta))^n$ to its right-hand side, and then put $\alpha = \beta = 1$ in the result. We thus obtain the more general identity

$$\int_0^\infty \exp[-y^2/2] y^{\nu+1/2} T_n^{(\nu)}(y^2) \sqrt{(xy)} \,.\, J_\nu(xy)\, dy =$$

$$= (-1)^n \exp[-x^2/2] x^{\nu+1/2} T_n^{(\nu)}(x^2), \quad (16)$$

where $T_n^{(\nu)}(z)$ is a polynomial of degree n in z, and in particular,

$$T_0^{(\nu)}(z) = 1, \qquad T_1^{(\nu)}(z) = -z + (\nu+1),$$
$$T_2^{(\nu)}(z) = z^2 - 2(\nu+2)z + (\nu+1)^2.$$

We see that

$$\exp[-x^2/2] x^{\nu+1/2} T_n^{(\nu)}(x^2) \qquad (n = 0, 1, 2, \ldots)$$

is the eigenfunction corresponding to the eigenvalue $(-1)^n$ of the Hankel operator.

Since the Hankel operator is symmetric, the polynomials $T_n^{(\nu)}(t)$ and $T_m^{(\nu)}(t)$ with degrees of different parity satisfy the following orthogonality relation:

$$\int_0^\infty \exp[-y^2] y^{2\nu+1} T_n^{(\nu)}(y^2) T_m^{(\nu)}(y^2)\, dy = 0;$$

in other words, the polynomials $T_n^{(\nu)}(t)$ and $T_m^{(\nu)}(t)$ are orthogonal relative to the weight-function $e^{-t}t^\nu$ $(t>0)$:

$$\int_0^\infty e^{-t} t^\nu T_n^{(\nu)}(t) T_m^{(\nu)}(t)\, dt = 0.$$

This is the situation as regards polynomials having different parity. By orthogonalizing each of the two systems

$$\{T_{2k}^{(\nu)}(t)\}_{k=0}^\infty, \qquad \{T_{2k+1}^{(\nu)}(t)\}_{k=0}^\infty$$

and combining them, we obtain the system of polynomials

$$L_n^{(\nu)}(t) = (-1)^n \frac{1}{n!} t^n + \cdots \qquad (n = 0, 1, 2, \ldots)$$

such that

$$\int_0^\infty e^{-t} t^\nu L_n^{(\nu)}(t) L_m^{(\nu)}(t)\, dt = 0 \qquad (m \neq n).$$

This orthogonality relation is clearly equivalent to the following relation:

$$\int_0^\infty e^{-t} t^\nu L_n^{(\nu)}(t) t^m\, dt = 0 \qquad (n>0;\ m = 0, 1, \ldots, n-1). \quad (17)$$

The name given to these polynomials reflects the fact that a system of this type, with $\nu = 0$, was first constructed by Laguerre,† and later, for any $\nu > -1$, by N. Sonine.

The relation (17) determines the orthogonal polynomials up to a constant factor, but since the leading coefficient of the polynomial $L_n^{(\nu)}(t)$ has already been chosen, we can easily show that

$$L_n^{(\nu)}(t) = \frac{1}{n!} e^t t^{-\nu} \frac{d^n}{dt^n} (e^{-t} t^{n+\nu}) \qquad (n = 0, 1, \ldots);$$

because, with this expression for $L_n^{(\nu)}(t)$, the orthogonality relations (17) are easily verified by integration by parts. This explicit formula for $L_n^{(\nu)}(t)$

† We have already encountered this system in §12, but with a different normalization.

also enables us to find the *normalization coefficient:*

$$a_n^2 = \int_0^\infty e^{-t}t^\nu [L_n^{(\nu)}(t)]^2\,dt = \int_0^\infty e^{-t}t^\nu L_n^{(\nu)}(t)\frac{(-1)^n t^n}{n!}\,dt$$

$$= \frac{(-1)^n}{(n!)^2}\int_0^\infty t^n \frac{d^n}{dt^n}(e^{-t}t^{n+\nu})\,dt = \frac{n!}{(n!)^2}\int_0^\infty e^{-t}t^{n+\nu}\,dt = \frac{\Gamma(n+\nu+1)}{\Gamma(n+1)}.$$

Let us prove that in the space L^2 with the weight $e^{-t}t^\nu$ $(0<t<\infty)$ the orthogonal sequence $\{L_n^{(\nu)}(t)\}_{k=0}^\infty$ is complete. Assuming the contrary, suppose that there is a measurable function $\Phi(t)$ $(0<t<\infty)$ such that

$$\int_0^\infty e^{-t}t^\nu\,|\Phi(t)|^2\,dt \neq 0$$

and that nevertheless

$$\int_0^\infty e^{-t}t^\nu \Phi(t) L_n^{(\nu)}(t)\,dt = 0 \qquad (n=0, 1, 2, \ldots)$$

or, equivalently, that

$$\int_0^\infty e^{-t}t^\nu \Phi(t) t^n\,dt = 0 \qquad (n=0, 1, 2, \ldots).$$

Repeating here essentially the argument already used at the end of §12, we consider the function

$$g(\zeta) = \int_0^\infty e^{-t}t^\nu \Phi(t) e^{-it\zeta}\,dt \qquad (\zeta = \xi + i\eta)$$

which, clearly, is regular in the half-plane $\eta < 1/4$, and for which, by our assumption,

$$g^{(n)}(0) = \int_0^\infty e^{-t}t^\nu \Phi(t)(-it)^n\,dt = 0 \qquad (n=0, 1, 2, \ldots).$$

But in that case $g(\zeta)\equiv 0$ and therefore

$$g(\xi) = \int_0^\infty e^{-t}t^\nu \Phi(t) e^{-it\xi}\,dt \equiv 0 \qquad (-\infty<\xi<\infty);$$

this means that the Fourier transform is identically zero of a function which is 0 when $t<0$, and which, when $t>0$, is $e^{-t}t^\nu\Phi(t)\in L^1(0,\infty)$. Therefore $\Phi(t)=0$, and we have obtained a contradiction.

We shall now prove the Hankel inversion formulae. The Hankel operator on a function $f(x)$ makes sense immediately if

$$\int_0^\infty |f(x)|\,dx<\infty \qquad \text{and} \qquad \int_0^1 |f(x)|\,x^{\nu+1/2}\,dx<\infty.$$

The second condition is certainly satisfied if

$$\int_0^1 |f(x)|^2\,dx<\infty.$$

However, we must assume only that $f\in L^2(0,\infty)$. But in this case we can expand $f(x)$ in a series of eigenfunctions of the operator H, *i.e.*, in a series which converges in the metric of $L^2(0,\infty)$:

$$f(x)\sim\sum_{k=0}^{\infty} c_k \exp[-x^2/2]x^{\nu+1/2}L_k^{(\nu)}(x^2).$$

On the parts

$$f_n(x)=\sum_{k=0}^{n} c_k \exp[-x^2/2]x^{\nu+1/2}L_k^{(\nu)}(x^2)$$

of this series the Hankel operator can be defined directly:

$$g_n(x)=(Hf_n)(x)=\sum_{k=0}^{n}(-1)^k c_k \exp[-x^2/2]x^{\nu+1/2}L_k^{(\nu)}(x^2).$$

But $\{g_n(x)\}_0^\infty$ is a Cauchy sequence, and therefore converges in $L^2(0,\infty)$ to some function $g(x)$, which has the expansion

$$g(x)\sim\sum_{k=0}^{\infty}(-1)^k c_k \exp[-x^2/2]x^{\nu+1/2}L_k^{(\nu)}(x^2).$$

The operator H is unitary and self-adjoint. It can also be defined by the formula

$$(Hf)(x)=\underset{A\to\infty}{\text{l.i.m.}}\int_0^A f(y)\sqrt{xy}\,J_\nu(xy)\,dy.$$

The Chebyshev–Hermite polynomials are included as a particular case in the system of Laguerre–Sonine polynomials, *viz.*,

$$H_{2m}(t)=(-1)^m 2^{2m} m!\,L_m^{(-1/2)}(t^2),$$

$$H_{2m+1}(t)=(-1)^m 2^{2m+1} m!\,tL_m^{(1/2)}(t^2).$$

We may also mention that the Laguere–Sonine polynomials are connected with certain second-order differential expressions, and the rôle which these polynomials play in many topics of theoretical physics is due to this fact. The most important of these differential expressions has the form (we can, of course, assume that $\nu \neq 1/4$)

$$l = -\frac{d^2}{dt^2} + \left(t^2 + \frac{\nu^2 - 1/4}{t^2}\right) \qquad (0 < t < \infty).$$

The operators L^+ and L^- generated by this differential expression on the intervals $[1, \infty)$ and $(0, 1]$ have the deficiency index $(1, 1)$ (see Theorem 5 in §128). Therefore the deficiency index of the operator L on the interval $(0, \infty)$ is $(0, 0)$, *i.e.*, L is a self-adjoint operator. Since the equation $l[u] - \lambda u = 0$ is satisfied, when $\lambda = 4n + 2(\nu + 1)$ $(n = 0, 1, 2, \ldots)$, by the sequence of Laguerre–Sonine functions

$$\varphi_n^{(\nu)}(t) = \exp[-t^2/2]t^{\nu+1/2}L_n^{(\nu)}(t^2) \qquad (n = 0, 1, 2, \ldots)$$

which forms a complete system in $L^2(0, \infty)$, it follows that the operator L has a purely discrete spectrum here, just as in the Chebyshev–Hermite case (see example 4 of this section); and the inversion formulae relating to the operator L reduce to the expansions in terms of the functions $\varphi_n^{(\nu)}(t)$.

References

1. Akhiezer, N. I. On the theory of maximal symmetric operators in Hilbert space. *Uchen. Zap. Kharkov Aviats. Inst.*, **3,** no. 2, 1940 (in Russian).
2. Akhiezer, N. I. On some inversion formulae for singular integrals. *Izvest. Akad. Nauk SSSR Ser. Mat.*, **9,** 1945 (in Russian).
3. Akhiezer, N. I. Integral operators with Carleman kernals. *Uspekhi Mat. Nauk*, **2,** no. 5 (21), 1947 (in Russian).
4. Akhiezer, N. I. On orthogonal polynomials on several intervals. *Dokl. Akad. Nauk SSSR*, **134,** 1960 (in Russian).
5. Akhiezer, N. I. Continuous analogues of orthogonal polynomials on a system of intervals. *Dokl. Akad. Nauk SSSR*, **141,** 1961 (in Russian).
6. Akhiezer, N. I. A Sturm–Liouville equation on the half-line. *Zapiski Kharkov Mat. Obshch.*, **29,** 1965 (in Russian).
7. Akhiezer, N. I. On a Lamé equation. *Zapiski Kharkov Mat. Obshch.*, **30,** 1964 (in Russian).
8. Akhiezer, N. I. *The classical problem of moments*, Fizmatgiz, Moscow, 1961 (in Russian). English translation published by Oliver and Boyd, London and Edinburgh, 1965.
9. Akhiezer, N. I. *Lectures on the theory of approximations*, 2nd edn, Nauka, Moscow, 1965 (in Russian). English translation published by Ungar, New York, 1956.
10. Allakhverdiev, J. On the rate of approximation to completely continuous operators by finite-dimensional operators. *Uchen. Zap. Azerb. Univ.*, **2,** 1957 (in Russian).
11. Aronszain, N. and Smith, K. T. Invariant sub-spaces of completely continuous operators. *Ann. Math.* (2), **60,** 345–350, 1954.
12. Bateman, H. *Higher transcendental functions*, Vol. I, McGraw-Hill, New York, 1953.
13. Berezansky, Yu. M. Eigenfunction expansions for second-order partial difference equations. *Trudy Mosk. Mat. Obshch.*, 203–268, 1956 (in Russian).

14. Berezansky, Yu. M. *Eigenfunction expansions from self-adjoint operators*, Naukova Dumka, Kiev, 1965 (in Russian). English translation published by American Mathematical Society, Transl. of Math. Monograph **17,** Providence, R.I., 1968.
15. Bernstein, A. and Robinson, A. Solution of an invariant sub-space problem of K. Smith and P. Halmos. *Pacific J. Math.*, **16,** 421–432, 1960.
16. Birman, M. Sh. On existence conditions for wave operators. *Dokl. Akad. Nauk SSSR*, **143,** no. 3, 506–509, 1962 (in Russian). English transl. *Soviet Math. Dokl.*, **3,** 408, 1962.
17. Birman, M. Sh. A local condition for the existence of wave operators. *Dokl. Akad. Nauk SSSR*, 485–488, 1964 (in Russian). English transl. *Soviet Math. Dokl.*, **5,** 1505, 1964.
18. Birman, M. Sh. and Krein, M. G. On the theory of wave operators and scattering operators. *Dokl. Akad. Nauk SSSR*, **144,** no. 3, 476–478 (in Russian). English transl. *Soviet Math. Dokl.*, **3,** 704–747, 1962.
19. Brodsky, M. S. On triangular representations of completely continuous operators with a single point spectrum. *Uspekhi Mat. Nauk*, **16,** no. 1, 135–141, 1961 (in Russian).
20. Brodsky, M. S. and Livshits, M. S. Spectral analysis of non-self-adjoint operators and intermediate systems. *Uspekhi Mat. Nauk*, **13,** no. 1, 3–85, 1958 (in Russian). English transl. *Amer. Math, Soc. Transl.* (2), **13,** 265–346.
21. Brouwer, L. E. J. On one-to-one continuous transformations of surfaces into themselves. *Math. Ann.*, **69,** 176–180, 1910 (in German).
22. Carleman, T. *On singular integral equations with real symmetric kernals*, Almquist and Wiksells, Uppsala, 1923 (in French).
23. Cronin, J. Fixed points and topological degree in non-linear analysis. *Math. Surveys*, no. 11, 76–83, 1967.
24. Dunford, N. and Schwartz, J. T. *Linear operators*, Vol. 1, Interscience Publishers, New York, 1958.
25. Enflow, P. A counter-example to the approximation problem in Banach spaces. *Acta Math.*, **130,** 309–317, 1973.
26. Everitt, W. N. Integrable-square solutions of ordinary differential equations. *Quart. J. Math., Oxford* (2), **10,** 145–155, 1959.
27. Everitt, W. N. Singular differential equations. I. The even-order case. *Math. Ann.*, **156,** 9–24, 1964.
28. Freudenthal, H. On the Friedrich extension of semi-bounded operators. *Akad. van Wetensch Proc.*, **39,** no. 7, 832–833, 1936 (in German).
29. Friedrichs, K. O. Spectral theory of semi-bounded operators and

application to the spectral resolution of differential operators. II. *Math. Ann.*, **109,** 685–713, 1934 (in German).

30. Friedrichs, K. O. On the exceptional boundary-condition in the spectral theory of semi-bounded ordinary differential operators of the second order. *Math. Ann.*, **112,** 1–23, 1935 (in German).
31. Gasymov, M. G. and Levitan, B. M. Determination of a differential operator from two of its spectra. *Uspekhi Mat. Nauk*, **19,** no. 2, 3–63, 1964 (in Russian).
32. Gelfand, I. M. A lemma in the theory of linear spaces. *Soobshch. Kharkov. Mat. Obshch., Ser. 4*, **13,** 35–40, 1934 (in Russian).
33. Gelfand, I. M. and Shilov, G. E. *Generalized functions*, Vol. 2, *Some topics in the theory of differential equations*, Fizmatgiz, Moscow, 1958 (in Russian). English translation published by Academic Press, New York and London, 1967.
34. Gelfand, I. M. and Vilenkin, N. Ya. *Generalized functions*, Vol. 4, *Some applications of harmonic analysis; rigged Hilbert spaces*, Fizmatgiz, Moscow, 1961 (in Russian). English translation published by Academic Press, New York and London, 1968.
35. Glazman, I. M. On deficiency indices of differential operators. *Dokl. Akad. Nauk SSSR*, **44,** no. 2, 151–152, 1949 (in Russian).
36. Glazman, I. M. On the theory of singular quasi-differential operators. *Uspekhi Mat. Nauk*, no. 6, 102–135, 1950 (in Russian).
37. Glazman, I. M. *Direct methods of qualitative spectral analysis of singular differential operators*, Fizmatgiz, Moscow, 1963 (in Russian). English translation published by Israel Program for Scientific Translation, Jerusalem, 1965 and by Daniel Davey and Co., New York, 1966.
38. Glazman, I. M. and Naiman, P. B. On the convex hull of orthogonal functions. *Dokl. Akad. Nauk SSSR*, **102,** no. 3, 1955 (in Russian).
39. Gokhberg, I. Ts. and Krein, M. G. *Introduction to the theory of linear non-self-adjoint operators in Hilbert space*, Nauka, Moscow, 1965 (in Russian). English translation published by American Mathematical Society, Transl. of Math. Monograph **18,** Providence, R. I., 1969.
40. Gokhberg, I. Ts. and Krein, M. G. *Theory of Volterra operators in Hilbert space and their applications*, Nauka, Moscow, 1967 (in Russian). English translation published by American Mathematical Society, Transl. of Math. Monograph **24,** Providence, R.I., 1970.
41. Graff, A. A. On the theory of linear differential operators in a one-dimensional domain. *Mat. Sbornik*, **18,** 302–328, 1946; **21,** 143–159, 1947 (in Russian).
42. Hamburger, H. L. Contributions to the theory of convergence of Stieltjes continued fractions. *Math. Z.*, **4,** 186–222, 1919 (in German).

43. Hartman, P. *Ordinary differential equations*, John Wiley, New York, 1964, Chap. XI, §9.
44. Hilbert, D. *Foundations of a general theory of linear integral equations*, Teubner, Leipzig, 1912 (in German).
45. Hille, E. Non-oscillation theorems. *Trans. Amer. Math. Soc.*, **64,** 234–252, 1948.
45a. Hille, E. and Phillips, R. S. *Functional analysis and semi-groups.* Amer. Math. Soc. Colloquium Publications, **31,** Providence, R.I., 1957.
45b. Hu Sze-Tsen, *Homotopy theory*, Academic Press, New York, 1959.
46. Kakutani, S. Topological properties of the unit sphere in Hilbert space. *Proc. Imp. Acad., Tokyo*, **19,** 1973.
47. Kato, T. On finite-dimensional perturbations of self-adjoint operators. *J. Math. Soc. Japan*, **9,** no. 2, 230–249, 1957.
48. Kato, T. Perturbations of continuous spectra by trace class operators. *Proc. Japan. Acad.*, **33,** no. 5, 260–264, 1957.
49. Kato, T. *Perturbation theory for linear operators*, Springer Verlag, New York, 1966.
50. Kato, T. Wave operators and unitary equivalence. *Pacific J. Math.*, **15,** 171–180, 1965.
51. Kats, I. S. On Hilbert spaces generated by monotonic Hermitian matrix-functions. *Zap. Nauchno-issled. Inst. Math. Kharkov Mat. Obshch.*, **22,** 95–114, 1950 (in Russian).
52. Kilpi, Y. On self-adjoint extensions of symmetric transformations in Hilbert space. *Ann. Acad. Sci. Fennicae*, *Ser.* A, **1,** no. 264, 1959 (in German).
53. Kogan, V. I. and Rofe-Beketov, F. S. On square-integrable solutions of symmetric systems of differential equations. *Proc. Roy. Soc. Edinburgh, Ser. A*, **74,** 5–40, 1976.
54. Kogan, V. I. and Rofe-Beketov, F. S. On the question of the deficiency indices of differential operators with complex coefficients. *Proc. Roy. Soc. Edinburgh, Ser. A*, **72,** no. 4, 281–298 (1975).
55. Kolmogorov, A. N. and Fomin, S. V. *Elements of the theory of functions and functional analysis*, 3rd edn, Moscow, 1972, pp. 328 and 332 (in Russian). English translation published by Graylock Press, Rochester, N.Y., 1957.
56. Korotkov, V. B. On integral operators with Carleman kernels. *Dokl. Akad. Nauk SSSR*, **165,** 748–751, 1965 (in Russian).
57. Krasnoselsky, M. A. On self-adjoint extensions of Hermitian operators. *Ukrain. Mat. Zh.*, **1,** no. 1, 21–38, 1949 (in Russian).
58. Krein, M. G. On the resolvents of a Hermitian operator with deficiency indices (m, m). *Dokl. Akad. Nauk SSSR*, **52,** no. 8, 657–660, 1946 (in Russian).

59. Krein, M. G. On a general method of resolving positive-definite kernels into elementary products. *Dokl. Akad. Nauk SSSR*, **53,** no. 1, 3–6, 1946 (in Russian).
60. Krein, M. G. Theory of self-adjoint extensions of semi-bounded Hermitian operators and its applications. *Mat. Sbornik*, **20,** 431–495, 1947; **21,** 366–404, 1947 (in Russian).
61. Krein, M. G. On Hermitian operators and direction functionals. *Akad. Nauk Ukrain. RSR*, no. 10, 83–106, 1948 (in Russian).
62. Krein, M. G. Solution of the inverse Sturm–Liouville problem. *Dokl. Akad. Nauk SSSR*, **76,** 21–24, 1951 (in Russian).
63. Krein, M. G. Determination of the density of an inhomogeneous symmetric string from its frequency spectrum. *Dokl. Akad. Nauk SSSR*, **76,** 345–348, 1951 (in Russian).
64. Krein, M. G. On inverse problems for inhomogeneous strings. *Dokl. Akad. Nauk SSSR*, **82,** 669–672, 1952 (in Russian).
65. Krein, M. G. On a generalization of Stieltjes's investigations. *Dokl. Akad. Nauk SSSR*, **87,** 881–884, 1952 (in Russian).
66. Krein, M. G. On the transition function for a one-dimensional second-order boundary-value problem. *Dokl. Akad. Nauk SSSR*, **88,** 5–8, 1953 (in Russian).
67. Krein, M. G. and Krasnoselsky, M. A. Fundamental theorems on the extension of Hermitian operators and some applications to the theory of orthogonal polynomials and the problem of moments. *Uspekhi Mat. Nauk*, **2,** no. 3, 60–107, 1947 (in Russian).
68. Krein, M. G., Krasnoselsky, M. A. and Milman, D. P. On the deficiency indices of linear operators in Banach spaces and some geometrical topics. *Sbornik Trudov Inst. Mat. Akad. Nauk Ukrain. RSR*, **11,** 97–112, 1948 (in Russian).
69. Lax, P. D. Integrals of non-linear equations of evolution and soliton waves. *Commun. Pure Appl. Math.*, **21,** 467–490, 1968.
70. Levitan, B. M. *Eigenfunction expansions*, Gostekhizdat, Moscow, 1950 (in Russian).
71. Levitan, B. M. *Almost-periodic functions*, Gostekhizdat, Moscow, 1953 (in Russian).
72. Levitan, B. M. and Sargsyan, I. S. *Introduction to spectral theory; self-adjoint ordinary differential operators*, Nauka, Moscow, 1970 (in Russian). English translation published by American Mathematical Society, Transl. of Math. Monograph **39,** Providence, R.I., 1975.
73. Lewin, S. Some properties connected with convergence in the mean of sequences of functions. *Math. Z.*, **32,** 1930 (in German).
74. Livshits, M. S. On an application of the theory of Hermitian operators to the generalized problem of moments. *Dokl. Akad. Nauk SSSR*, **44,** 1944 (in Russian).

75. Livshits, M. S. On a class of linear operators in Hilbert space. *Mat. Sbornik*, **19,** 239–260, 1946 (in Russian). English translation in *Amer. Math. Soc. Transl.* (*2*), **13,** 61–83.
76. Livshits, M. S. On the theory of isometric operators with equal deficiency indices. *Dokl. Akad. Nauk SSSR*, **58,** 13–15, 1947 (in Russian).
77. Livshits, M. S. On the spectral resolution of linear non-self-adjoint operators. *Mat. Sbornik*, **34,** no. 1, 145–198, 1954 (in Russian). English translation in *Amer. Math. Soc. Transl.* (*2*), **5,** 67–114.
78. Lomonosov, V. I. On invariant subspaces of a family of operators which commute with a completely continuous operator. *Funktsion. analiz i ego prilozhenie*, **7,** no. 3, 55–56, 1973 (in Russian).
79. Lyusternik, L. A. and Sobolev, V. I. *Elements of functional analysis*, Nauka, Moscow, 1965 (in Russian). English translation published by Hindustan Publishing Corporation, New Dehli and John Wiley, New York, 1961.
80. Maclane, S. *Homology*, Grundlehren der mathematische Wissenschaft 114, Springer Verlag, Berlin, Heidelberg and New York, 1963.
81. Marchenko, V. A. Some topics in the theory of second-order differential operators. *Dokl. Akad. Nauk SSSR*, **72,** no. 3, 457–460, 1950 (in Russian).
82. Marchenko, V. A. Some topics in the theory of one-dimensional, linear, second-order differential operators. *Moscow. Math. Soc.*, **1,** 327–420, 1952; **2,** 3–89, 1953 (in Russian).
83. Marchenko, V. A. *Spectral theory of Sturm–Liouville operators*, Kiev, 1972 (in Russian).
84. Møller, C. General properties of the characteristic matrix in the theory of elementary particles. *Dan. Vid. Selsk. Fys. Medd.*, **21,** no. 1, 1945; **22,** no. 19, 1946.
85. Naimark, M. A. On a self-adjoint extension of the second kind of a symmetric operator. *Isvest. Akad. Nauk SSSR, Ser. Mat.*, **4,** no. 1, 53–104, 1940 (in Russian).
86. Naimark, M. A. Spectral functions of a symmetric operator. *Isvest. Akad. Nauk SSSR, Ser. Mat.*, **4,** 277–318, 1940 (in Russian).
87. Naimark, M. A., On a representation of additive operator set-functions. *Dokl. Akad. Nauk SSSR*, **41,** 359–361, 1943 (in Russian).
88. Naimark, M. A. On spectral functions of a symmetric operator. *Isvest. Akad. Nauk SSSR*, **1,** no. 6, 285–296, 1943 (in Russian).
89. Naimark, M. A. Extremal spectral functions of symmetric operators. *Isvest. Akad. Nauk SSSR*, **11,** no. 4, 1947 (in Russian).

90. Naimark, M. A. *Linear differential operators*, Nauka, Moscow, 1969 (in Russian). English translation published by Ungar, New York, 1968.

91. von Neumann, J. On the theory of unbounded matrices. *Z. reine angew. Math.*, **16,** 1929 (in German).

92. von Neumann, J. General theory of eigenvalues of Hermitian functional operators. *Math. Ann.*, **102,** 49–131, 1929–30 (in German).

93. von Neumann, J. On functions of function-operators. *Ann. Math.* (*2*), **32,** 191–226, 1931 (in German).

94. von Neumann, J. On a theorem of Mr M. H. Stone. *Ann. Math.* (*2*), **33,** 567–573, 1932 (in German).

95. von Neumann, J. Characterization of the spectrum of an integral operator. *Acta Sci. Ind., Paris*, **229,** 1935 (in German).

96. Newton, R. G. *Scattering theory of waves and particles*, McGraw-Hill, New York, 1966, Chap. 4, §7.

97. Paley, R. E. A. C. and Wiener, N. *Fourier transforms in the complex domain*, American Mathematical Society Colloquium Publications 19, American Mathematical Society, New York, 1934.

98. Pichorides, S. K. On the best possible values of the constants in the theorems of M. Riesz, Zygmund and Kolmogorov. *Studia Math.*, **44,** 165–179, 1972.

99. Plesner, A. I. Spectral theory of maximal operators. *Dokl. Akad. Nauk SSSR*, **22,** no. 5, 227–230, 1939 (in Russian).

100. Plesner, A. I. *Spectral theory of linear operators*, Nauka, Moscow, 1965 (in Russian). English translation published by Ungar, New York, 1969.

101. Povzner, A. Ya. On M. G. Krein's method of direction functions. *Zap. nauchno-issled. Inst. Mat. Mekh. Kharkov. Mat. Obshch., Ser. 4*, **20,** 43–52, 1950 (in Russian).

102. Privalov, I. I. *Limit properties of analytic functions*, Gostekhizdat, Moscow, 1956 (in Russian).

103. Riesz, F. On linear functional equations. *Acta Math.*, **41,** 71–98, 1918 (in German).

104. Riesz, F. On functions of Hermitian transformations in Hilbert space. *Acta Sci. Math., Szeged*, **7,** 147–159, 1935 (in French).

105. Riesz, F. and Sz-Nagy, B. *Lessons of functional analysis*, Budapest, 1953 (in French). English translation published by Ungar, New York, 1955.

106. Rofe-Beketov, F. S. Eigenfunction expansions for infinite systems of differential equations. *Trudy 5th Vse-Soyuz. Konf. funkts. anal. i ego pril., Baku*, 1961, 230–237 (in Russian).

107. Rofe-Beketov, F. S. A criterion for the finiteness of the number of

discrete levels introduced in the lacunae of the continuous spectrum by perturbations of a periodic potential. *Dokl. Akad. Nauk SSSR*, **156,** no. 3, 515–518, 1964 (in Russian).

108. Rofe-Beketov, F. S. Self-adjoint extensions of differential operators in a space of vector functions. *Dokl. Akad. Nauk SSR*, **184,** no. 5, 1969 (in Russian).
109. Rofe-Beketov, F. S. Self-adjoint extensions of differential operators in a space of vector functions. *Teor. funkts., funkts anal. ego pril.*, **8,** 3–24, 1969 (in Russian).
110. Rofe-Beketov, F. S. Deficiency indices and properties of the spectrum of some classes of differential operators. In *Spectral theory and differential equations,* Lecture Notes in Mathematics 448, Springer Verlag, Berlin, Heidelberg and New York, 273–293 1974.
111. Rosenblum, M. Perturbations of the continuous spectrum and unitary equivalence. *Pacific J. Math.*, **7,** no. 1, 997–1010, 1967.
112. Schauder, J. On the theory of continuous mappings in function space. *Math. Z.*, **26,** 1927 (in German).
113. Shilov, G. E. and Gurevich, B. L. *Integral, measure, and derivative,* Nauka, Moscow, 1964 (in Russian). English translation published by Prentice-Hall, Englewood Cliffs, N.J., 1966.
114. Shtraus, A. B. Generalized resolvents of symmetric operators. *Izvest. Akad. Nauk SSSR, Ser. Mat.*, **18,** 51–86, 1954 (in Russian).
115. Shtraus, A. B. A formula for generalized resolvents of a differential operator of even order. *Dokl. Akad. Nauk SSSR*, **111,** no. 4, 1956 (in Russian).
116. Shtraus, A. B. Characteristic functions of linear operators. *Izvest. Akad. Nauk SSSR, Ser. Mat.*, **24,** 43–74, 1960 (in Russian).
117. Shtraus, A. B. On extensions of a symmetric operator depending on a parameter, *Izvest. Akad. Nauk SSSR, Ser. Mat.*, **29,** 1389–1410, 1965 (in Russian).
118. Smirnov, V. I. *Course of higher mathematics,* Vol. 5, Fizmatgiz, Moscow, 1959 (in Russian). English translation published by Pergamon Press, Oxford, 1964.
119. Stone, M. H. *Linear transformations in Hilbert space and their applications to analysis,* American Mathematical Society Colloquium Publications **15,** American Mathematical Society, New York, 1932.
120. Sz.-Nagy, B. *Spectral representation of linear transformations of Hilbert space,* Ergebnisse der Mathematik 5, Springer Verlag, Berlin, Heidelberg and New York, 1942 and 1967 (in German). Also reprinted by Edwards Bros, Ann Arbor, Mich., 1947.
121. Sz.-Nagy, B. Perturbations of self-adjoint operators in Hilbert space. *Comment. Math. Helv.*, **19,** 347–366, 1947 (in French).

122. Sz.-Nagy, B. and Foiąs, C. *Harmonic analysis of operators in Hilbert space*, Akad. Kaido, Budapest, 1967 (in French).
123. Titchmarsh, E. C. *Introduction to the theory of the Fourier integral*, Oxford University Press, Oxford, 1937.
124. Titchmarsh, E. C. *Eigenfunction expansions associated with second-order differential equations*, Vols I and II, Oxford University Press, Oxford, 1962 and 1958.
125. Weyl, H. On bounded quadratic forms whose difference is completely continuous. *Rend. Circolo Mat. Palermo*, **27,** 373–392, 1909 (in German).
126. Weyl, H. On ordinary differential equations with singularities and the associated expansions of arbitrary functions. *Math. Ann.*, **68,** 220–269, 1910 (in German).
127. Whittaker, E. T. and Watson, G. N. *A course of modern analysis*, Cambridge University Press, Cambridge, 1962.

Index

Note: *pages* 1–312 *appear in Volume I and pages* 313–552 *appear in Volume II*